Are we there yet?

Are we there yet?
Case studies of implementing decision support for patients

Commentaries from three perspectives
Relational Coordination Theory **by Jody Hoffer Gittell**
Normalization Process Theory **by Glyn Elwyn**
Microsystems **by Marjorie M. Godfrey**

Edited by
Glyn Elwyn
Stuart W. Grande
Jody Hoffer Gittell
Marjorie M. Godfrey
Dale Collins Vidal

The Dartmouth Center for Health Care Delivery Science
The Dartmouth Institute for Health Policy and Clinical Practice
Hanover, New Hampshire

Copyright © 2013 by Trustees of Dartmouth College on behalf of The Dartmouth Center for
Health Care Delivery Science and The Dartmouth Institute for Health Policy and Clinical Practice
All rights reserved

Printed in the United States of America
Cover design by Benjamin Elwyn
Typeset by Linnea Spelman

First Edition

No part of this publication may be reproduced, stored in or introduced into a retrieval system,
or transmitted, in any form, or by any means, including electronic, mechanical, photocopying,
recording, or otherwise, without prior written permission, except in the case of brief quotations
embodied in critical reviews and certain other noncommercial uses permitted by copyright law.
The copyright in individual sections or chapters of this book is held as provided under applicable
law or as provided by agreement among the relevant persons or entities. See the end of each
section or chapter for the name of the person to contact regarding obtaining permission to use
such section or chapter.

ISBN 978-0-9894056-2-1

*This book is dedicated to patients everywhere,
with apologies that we are so slow in fixing healthcare.*

CONTENTS

Contents vii
List of Figures and Tables ix
Foreword xi
Preface xv
Acknowledgements xvii

IMPLEMENTATION PERSPECTIVES

Introduction. 3
Summaries of Three Implementation Perspectives
   2.1   Relational Coordination Theory 7
   2.2   Normalization Process Theory 10
   2.3   Microsystems 12

CASE STUDIES AND COMMENTARIES

Implementing Shared Decision Making in the
United Kingdom: Lessons from the Making Good
Decisions in Collaboration (MAGIC) Program. 23

Implementing Decision Support and Shared Decision Making:
Steps Toward Culture Change 35

Shared Decision Making at Group Health:
Distributing Patient Decision Support and Engaging Providers . . . 45

Step-by-step Using a Community Effort:
Implementing Patient Decision Support in Stillwater, Minnesota . . 57

Orientation and Education: Steps Needed
to Get Clinicians to Order Patient Decision Support 67

Implementing Tailored Decision Support
Using a Patient Health Survey. 77

Integration Challenges: Loss of an Existing Patient Decision
Support System When a New Electronic Health Record Arrives. . . . 87

CONTENTS

NEXT STEPS
Where Are We? Where Do We Go Next? 101
List of Authors and Editors. 117
Selected Glossary 121
Institute Team
    Co-Chairs. 125
    Management Team 126
    Faculty Advisory Committee 127
Summer Institute Participants, 2011 129

LIST OF FIGURES AND TABLES

FIGURE 2.1.1
Relational coordination and relational coproduction 8

FIGURE 2.1.2
Relational model of organizational change 9

FIGURE 2.3.1
Success characteristics of high performing clinical microsystems 13

FIGURE 2.3.2
Systems within systems in health care 14

FIGURE 2.3.3
Multiple microsystems create mesosystem 14

FIGURE 3.1
Logic model for the MAGIC program 26

FIGURE 3.2
Shared decision making: A model for clinical practice 27

FIGURE 3.3
Ask 3 questions campaign at the Cardiff site of the MAGIC program . . . 29

FIGURE 4.1
Examples of social marketing campaign items
with PMDM brand name and tag lines 38

FIGURE 4.2
Percent of eligible patients who received DESIs
about colorectal cancer screening and back pain[3] 39

TABLE 5.1
Summary of decision aids, treatment choices, and total distribution . . . 47

FIGURE 5.1
Monthly decision aid distribution by service line 49

LIST OF FIGURES AND TABLES

TABLE 5.2
Facilitators and barriers to effective
decision aid implementation – All service lines 51

TABLE 6.1
Delivery rates for selected conditions at Stillwater Medical Group . . . 60

FIGURE 7.1
Performance feedback comparing prescription rates across practices. . . 70

FIGURE 7.2
Training intervention significantly
increased number of decision aids ordered 71

FIGURE 8.1
Clinical information system delivery flow-chart 79

TABLE 8.1
Number and types of decision aids distributed 80

FIGURE 9.1
Proposed decision making process for total knee replacement 90

FIGURE 9.2
Decision support integration. 91

FIGURE 9.3
Decision quality, decision process, decision aid acceptability 93

TABLE 10.1
Barriers and facilitators to implementation. 105

FOREWORD

Decision support for patients

Health care systems in the United States and elsewhere face daunting challenges. In the developed world, aging populations and rising health care costs threaten public and private budgets, while in low- and middle-income countries rising expectations and remarkably uneven health system performance undermine social stability. Evidence that the supply of health care resources and providers' opinions are a powerful determinant of utilization raises the possibility that both underuse and overuse are as likely to be a problem in developing countries as in the most advanced ones[1].

Shared decision making has therefore emerged a beacon of hope for many[2]. Shared decision making not only offers the possibility of minimizing harm and maximizing benefits by making sure that treatments are aligned with patients goals, values, and preferences; it also offers perhaps the best hope of determining patients' and populations' true (well-informed) demand for services[3].

But as this volume makes clear—the gap between promise and reality remains deep and wide. The seven case studies presented in this book offer important and useful observations about what happened when leaders in the field of shared decision making worked with willing local partners to try to implement decision-support for patients on the ground in a diverse group of health care settings in the United States and United Kingdom. In each case, some success in the distribution of decision-aids to patients was achieved. In most settings, however, the decision-support tools were distributed to many fewer patients than could have benefited. In only a minority of the cases was there any effort to understand whether receiving one of the decision-aids led either to "shared decision making" or decisions that were better aligned with patients' preferences. For each case study, the authors report what they learned—and these insights offer useful material and insights for the next generation of pioneers.

The commentaries by Jody Hoffer Gittell, Marjorie Godfrey, and Glyn Elwyn help us understand why broad adoption of shared decision making will be so hard. Gittell's work on relational coordination suggests that implementing and sustaining organizational change requires relational interventions (so that professionals can test new ways of interacting), work process interventions (changing the workflows that currently support successful operations), and structural interventions (such as new performance measures and incentives). Godfrey brings her experience teaching and coaching health systems by applying microsystem theory. The critical insight here is that care is largely delivered by small, front-line practice units where an individual patient with a specific clinical issue interacts with a well-defined team to achieve (or not) the intended results through processes that can be mapped, measured, and improved (imagine diabetes care in a primary care practice). Finally, through the lens of Normalization Process Theory, Elwyn frames the work of change by identifying four key tasks that must be carried out for any major change to be successfully implemented in an organization: defining the new work; determining who does it; figuring out how the work will get done; and tracking the process changes and the outcomes achieved while reflecting on what further changes will be required to sustain the new work.

All three perspectives highlight the key challenge: implementing shared decision making requires what turns out to be a major change in current processes of care. And when those changes involve reframing professional roles, the barriers are even greater. Just how serious these barriers are emerges not only from the analyses of the case studies, but also from Glyn Elwyn's provocative closing chapter. He underscores a very real problem: health professionals see their primary responsibility as diagnosing biologically mediated problems (what's the matter with the patient) and prescribing and delivering treatments to fix those problems. By and large providers neither value, nor put much effort into, the challenge of understanding patients' goals, values and preferences and making sure that decisions reflect an understanding of what matters most to the patient. He suggests eight specific steps that might be taken to help accelerate progress—all of which are reasonable. But all focus largely on the provider side of the equation.

A bit to my surprise, the list does not mention the possibility of harnessing the energy of patients themselves as an additional, perhaps powerful, engine of change. While there is no question that the major barrier to successful reform lies in both current work processes and professional norms, we should also consider the power that patents and the public could bring to this work. My increasing interest in making sure patients are at the table as we redesign care comes not only from my experience including patients on the teams developing tools to track patient reported outcomes of care (where they are extremely helpful), but from my recent visit to the hemodialysis unit at Ryhov County Hospital, in Jönköping, Sweden. There I met Christian Farman and Britt-Mari Banck. In 2005, Farman's transplanted kidney failed and he had to go back on dialysis. He had done enough research to convince himself that self-dialysis was possible and that it might be both safer and more effective (more frequent, shorter treatments allowing better control of symptoms and blood levels). He asked Banck, his nurse, if she could help him regain control of his life by teaching him to do hemodialysis himself: remarkably, she agreed to do so. Others soon followed, and now 60% of the patients at the unit perform self-dialysis, on their own schedule. Early evidence on clinical outcomes appears promising, but there is no doubt that the patients have been empowered as partners in their own care.

This book makes clear that we have a long way to go to ensure that shared decision-making becomes a routine element of care. We will need all the help we can get—from policy makers, providers—and patients.

Elliott Fisher, Director
The Dartmouth Institute for Health Policy and Clinical Practice

References

1. Fisher ES, Wennberg JE. Health care quality, geographic variations, and the challenge of supply-sensitive care. *Perspectives in Biology and Medicine.* 2003;46(1):69–79.

2. Oshima Lee E, Emanuel EJ. Shared decision making to improve care and reduce costs. *The New England Journal of Medicine.* 2013;368(1):6–8.

3. Mulley AG. Inconvenient truths about supplier induced demand and unwarranted variation in medical practice. *BMJ.* 2009;339:b4073.

PREFACE

> *It ought to be remembered that there is nothing more difficult to take in hand, more perilous to conduct, or more uncertain in its success, than to take the lead in the introduction of a new order of things.*
>
> Niccolo Machiavelli, The Prince (1553)

In many ways, this book is about unreasonable people who know that something is wrong with the "order of things" and want to introduce a new way of doing. These people, and the broader community of which they are a part, know that medicine as practiced now and in the past is often lacking. Despite spectacular advances in science and technology, these people are all too aware of the disconnect that frequently exists between what is done and what might be "best done" when the balance of harms and benefits are carefully considered.

In their quest for improvement, these unreasonable people gather evidence from randomized trials and worry about the bias that occurs when evidence remains unpublished and unavailable. They design tools that attempt to tell a clear story about treatment options allowing for comparisons on issues that matter to patients, but are often left unsaid in the hurried ways that medicine is too often practiced. They study the chance of receiving treatments merely by virtue of location, investigating the tenfold difference or more that often exists in procedure rates—a difference that is difficult to explain using any logic of varying local need. These unreasonable people think the current systems are fundamentally flawed, and they work to find solutions.

The individuals in this book are some of these unreasonable people. They have studied how tools, designed to translate scientific evidence into information that is more easily understood, can help patients faced with tough choices decide what is best done. These are stories told by pioneers who believe that their work will eventually make a difference. As you will see, the rewards are slim, and the challenges—conflicting incentives, resistant cultures, and systems designed to reward opposing behaviors and outcomes—seem insurmountable. But here they are; they seem to keep at it. We think we know why. They have heard it said that progress depends on unreasonable people.

Glyn Elwyn

ACKNOWLEDGEMENTS

The editors thank Eric Weinberger for meticulous copy-editing, Aileen Lem for editing, layout skills, proof-reading and her mastery of Mendeley, Robin Paradis Montibello for her support on bibliographies, and Dawn Carey and Allison Hawke for their work on organizing and planning the dissemination of this publication. Cover design by Benjamin Elwyn http://dk8.co/ and typesetting by Linnea Spelman, Three Monkeys Design Works.

Are we there yet?

SECTION ONE
IMPLEMENTATION PERSPECTIVES

CHAPTER 1

Introduction

Glyn Elwyn

THE SUMMER INSTITUTES FOR INFORMED PATIENT CHOICE at Dartmouth College have been a regular watering hole for a number of researchers over many years. Initiated by Hilary Llewellyn-Thomas and aided by her long-standing colleague Annette O'Connor, the Summer Institutes were first supported by funding from the Agency for Healthcare Research and Quality (AHRQ) and the Informed Medical Decisions Foundation (IMDF), formerly the Foundation for Informed Medical Decision Making (FIMDM). Attended by delegates from many countries, the Summer Institutes provide an environment for participants to learn more about how to inform and involve patients in health care decisions. You will encounter a plethora of terms including: shared decision making, informed patient choice, patient participation, patient decision aids, patient decision support interventions and tools, decision coaches, and so on. Don't get worried; we provide a glossary for the uninitiated. We also want to reassure you that we view this as a sign of a healthy debate in a new and exciting field of research. The terminology will eventually settle down[1].

Doing shared decision making is not the same as making sure patients gain access to tools—so-called patient decision support tools, aids, or interventions. There are many definitions of shared decision making, all of which have at their common core the communication process that occurs when clinical providers work with patients to arrive at good decisions based on up-to-date information (i.e., evidence), and where the preferences of patients and their providers are, to some extent, explicit and considered. There is much written about the nuances and details of how this might be best done. However, it is sufficient to say here that shared decision making is not about developing or implementing the tools used to support this process, but accepting that these tools are designed to be helpful adjuncts. In that sense, this set of case studies is not about shared decision making—it is about the work done to support the goal of putting patient decision support tools into practice.

A key resource for this field is the Cochrane systematic review of decision aids, led for many years by Annette O'Connor and now by Dawn Stacey[2]. By now the conclusions are well known: there is consistent evidence that when these interventions are delivered in controlled research settings, good things happen. Patient knowledge increases, as one would expect, from the delivery of high quality information. Patient understanding of risk probabilities improves and their reported participation in decision making increases. Reports about the impact on cost and on a measure known as "decision quality" are less clear and remain subject to conjecture. But there is no doubt that this systematic review provides an excellent foundation for future progress.

Over the many years during which these randomized controlled trials were being conducted, another pattern became evident. Although good patient outcomes were reported, there was no evidence that these interventions were being used routinely in clinical settings. The tools were developed, trials done, results analyzed, and papers published, yet clinicians did not continue using the interventions. Often the tools, once developed, were not subsequently available; nor were they updated. Some developers managed to sustain the tools for a number of years, but eventually when research grants were finished, the tools often disappeared. A research-practice gap became increasingly obvious. All of this led to the theme for the Summer Institute of 2011 being designated as "implementation," and detailing how to facilitate embedding these interventions into routine practice.

In the US, two organizations have managed to partially overcome some of these adoption hurdles, or so it seems. Healthwise, a not-for-profit organization, has successfully developed patient information tools and marketed them to a large number of health care organizations. Their tools include a significant number of text and web-based "decision aids." (Healthwise is the first to admit that these are less sophisticated tools than those used in many trials, and that they have not yet conducted evaluations about their effectiveness.)

The other significant organization is Health Dialog, responsible for the large-scale production and distribution of tools developed over many years by the Boston based Informed Medical Decisions Foundation (IMDF). By offering a range of services to providers, including the identification of at-risk patients, decision coaching, and decision aids (DVDs and booklets), Health Dialog, a for-profit company, has succeeded in creating significant market value with revenue in the form of royalties paid to IMDF. As a result, IMDF, for many years, has sufficient income to award a number of research grants as well as to support a US-based dissemination network composed of providers investigating how best to implement these tools into routine practice. A number of the cases reported in this book formed part of this dissemination network and other companies have as a result investigated whether or not there is a viable marketplace for these types of tools, based on the intentions signaled in the Affordable Care Act[3]. IMDF also provided support for the Center for Shared Decision Making at Dartmouth-Hitchcock Medical Center.

By summer 2011, it was clear widespread adoption of patient decision support would be difficult. Tipping points had been heralded but had not materialized[4]. Conceptual analyses[5] and reviews of existing studies revealed that for these tools to become part of "normal" routines more would be required than just the availability of high quality decision support tools[6]. It was then that the Summer Institute took place. Hurricane Irene blew through New Hampshire, preventing some participants from attending because of the storm. But it was a good event. Jody Hoffer Gittell helped us view the cases through the lens of her ideas about relational coordination[7]. Marjorie M. Godfrey provided the microsystems lens[8], and I introduced the normalization process theory developed by Carl May[9]. The following chapters are descriptions of the cases written by the teams that attended, followed by commentaries that consider these three different perspectives on how to implement patient decision support into practice.

Please direct permission requests for Chapter 1 to Glyn Elwyn at glynelwyn@gmail.com

References

1. Edwards AG, Elwyn G, eds. *Shared Decision-making in Health Care: Achieving Evidence-based Patient Choice*. 2nd ed. Oxford: Oxford University Press; 2009:3–10.

2. Stacey D, Bennett CL, Barry MJ, Col NF, Eden KB, Holmes-Rovner M, Llewellyn-Thomas HA, Lyddiatt A, Légaré F, Thomson RG. Decision aids for people facing health treatment or screening decisions. *Cochrane Database of Systematic Reviews*. 2011;10(10).

3. Senate and House of Representatives. *The Patient Protection and Affordable Care Act*. Washington: 111th Congress, 2nd Session; 2010.

4. O'Connor AM, Wennberg JE, Légaré F, Llewellyn-Thomas HA, Moulton BW, Sepucha KR, Sodano AG, King JS. Toward the "tipping point": Decision aids and informed patient choice. *Health Affairs*. 2007;26(3):716–725.

5. Elwyn G, Légaré F, Van der Weijden T, Edwards AG, May CR. Arduous implementation: Does the Normalisation Process Model explain why it's so difficult to embed decision support technologies for patients in routine clinical practice. *Implementation Science*. 2008;3(1):57.

6. Légaré F, Ratté S, Gravel K, Graham ID. Barriers and facilitators to implementing shared decision-making in clinical practice: Update of a systematic review of health professionals' perceptions. *Patient Education and Counseling*. 2008;73(3):526–535.

7. Gittell JH. Relational coordination: Guidelines for theory, measurement and analysis. *Relational Coordination Research Collaborative*. 2009.

8. Nelson EC, Godfrey MM, Batalden PB, Berry SA, Bothe AE, McKinley KE, Melin CN, Muething SE, Moore LG, Wasson JH, Nolan TW. Clinical microsystems, part 1. The building blocks of health systems. *Joint Commission Journal on Quality and Patient Safety / Joint Commission Resources*. 2008;34(7):367–78.

9. May CR, Mair F, Finch T, MacFarlane A, Dowrick C, Treweek S, Rapley T, Ballini L, Ong BN, Rogers A, Murray E, Elwyn G, Légaré F, Gunn J, Montori VM. Development of a theory of implementation and integration: Normalization process theory. *Implementation Science*. 2009;4(1):29.

CHAPTER 2

Summaries of Three Implementation Perspectives

2.1 Relational Coordination Theory
Jody Hoffer Gittell

RELATIONAL COORDINATION (RC) is defined as "communicating and relating for the purpose of task integration" or more simply "coordinating work through relationships of shared goals, shared knowledge and mutual respect." This concept first emerged in a study of flight departures[1-4] and its applicability to patient care quickly became apparent[5-12]. Relational coordination is a network of communication and relationship ties among workgroups that are engaged in a common task—for example, flight departures, patient care, or more specifically the discharge of patients from the operating room to the intensive care unit. It is a validated measure based on seven survey questions, including three about relationships: shared goals, shared knowledge, mutual respect, and four about communication: frequency, timeliness, accuracy, problem-solving.

The evidence thus far suggests that relational coordination improves quality and efficiency performance because it enables people to better manage their interdependencies with fewer dropped balls and less wasted effort. RC also improves job satisfaction and other worker outcomes by enabling people to be resilient under stress[13-16]. RC promotes both reciprocal learning and learning from failure[17,18]. Moreover the benefits of RC are expected to increase in the face of reciprocal task interdependence, uncertainty and time constraints[19].

The evidence also suggests that organizations can either promote or undermine relational coordination. The organizational structures that predict high levels of RC are those that connect across workgroups rather than reinforcing the silos that separate workgroups. The theory therefore calls for organizations to replace traditional bureaucratic structures with more relational structures—such as hiring and training for cross-functional teamwork, cross-functional conflict resolution, cross-functional performance measurement and rewards, cross-functional boundary spanners (like care managers or care coordinators), cross-functional meetings (like patient rounds), cross-functional protocols (like clinical pathways), and cross-functional information systems[3,7,10,20].

However to achieve desired performance outcomes, relational coordination needs to be extended beyond the team of care providers to include the patient and family in the process of shared decision making. We call engaging with the patient and family as members of the care team "relational coproduction." Like relational coordination, relational coproduction is characterized by relationships of shared goals, shared knowledge, and mutual respect, supported by frequent, timely, accurate, and problem-solving communication (see Figure 2.1.1)[20-22].

FIGURE 2.1.1
Relational coordination and relational coproduction

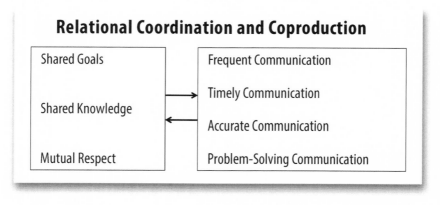

Together, relational coordination and relational coproduction are essential for achieving patient-centered care. But neither relational coordination nor relational coproduction is easy to achieve in the health care setting. We have learned that some organizational structures—those aforementioned cross-functional approaches to hiring, training, conflict resolution, rewards, performance measures, boundary spanner roles, meetings, protocols, and information systems—support relational coordination and relational coproduction. However, existing communication and relationship patterns in medicine are hierarchical and deeply embedded in professional identities and organizational cultures. Changing structures is not likely to be sufficient or even possible without addressing these deeper dynamics or long-established cultures.

In order to transform these deeply embedded patterns of interaction and sustain those structural changes, three types of interventions are needed (see Figure 2.1.2)

1. Relational interventions, or creating safe spaces to try out new ways of interacting among health care professionals;
2. Work process interventions, or improving work processes through structured problem-solving approaches like Plan-Do-Study-Act (PDSA) or lean management;
3. Structural interventions, or redesigning structures like hiring, training, conflict resolution, performance measurement, rewards, boundary spanner roles, meetings, protocols, information systems, etc., to support the desired new ways of interacting.

FIGURE 2.1.2
Relational model of organizational change

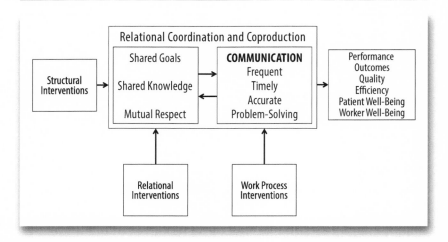

All the efforts documented in this book have used some of these interventions—structural, relational, work process—to foster higher levels of shared decision making between care providers and patients. My commentary throughout the book will use this framework to understand what has been done so far and which additional efforts might make sense going forward.

Please direct permission requests for Chapter 2 Section 1 to Jody Hoffer Gittell at jodyhoffergittell@brandeis.edu

2.2 Normalization Process Theory
Glyn Elwyn

THE NORMALIZATION PROCESS THEORY (NPT) developed by Carl May and colleagues focuses attention on factors that have been empirically demonstrated to affect the implementation and integration of complex interventions into health care organizations[23]. NPT has since undergone considerable development. The following brief descriptions are based on excellent summaries from the publically available website: www.normalizationprocess.org.

Identifying and adopting an innovative health technology, or a new way of organizing professional endeavors, is merely the start of implementation work. The task is really to make it normal, standard for the workplace adopting the new technology or processes for routine care, and it requires attention to many different issues. Policy-makers, managers, professionals, and patients all face two important problems as they try putting innovations into practice:

- **Process problems:** about the *implementation* of new ways of thinking, acting and organizing in health care;
- **Structural problems:** about the *integration* of new systems of practice into existing organizational and professional settings.

These are important problems for researchers and evaluators as well. To understand implementation and integration, we need to focus on the dynamic processes that lead to innovations becoming embedded in every day work. Normalization Process Theory facilitates understanding from a process evaluation perspective, with "normalization" defined as the routine embedding of a complex intervention in health care work, and NPT itself offering a robust structure for investigating the collective work that leads to implementation. NPT is structured along the following four constructs: *coherence, cognitive participation, collective action,* and *reflexive monitoring*.

Coherence
This is the sense-making work that people do individually and collectively when faced with the problem of operationalizing some set of practices. Implementation processes are chains of interactions in which a complex intervention (a new or modified way of thinking, acting upon, or organizing practice) is made coherent and enacted in a health care setting. Implementation processes are managed and "owned" through behaviors that denote cognitive participation—what we might call active thinking—by health care professionals and other personnel, including patients. The key question is, What is the work? In

the context of this book, the work is the use of decision support by patients and the challenges posed by this innovation.

Cognitive Participation
A complex intervention is enacted through different kinds of interactional work among many individuals. This work may be highly structured (enacting a research protocol, for example) or diffuse (e.g., operationalizing a policy decision in a large organization). This is the relational work of people to build and sustain a community of practice around a new technology or complex intervention. New service interventions often flounder because individuals aren't invested in ensuring that they fit with the ways that different groups of professionals—and sometimes patients—define their possible contribution to them. This is the work of keeping the new practices in view and connecting them with the people who need to be doing the work. The key question is, Who does the work? In the context of this book, we will examine who among the health care professionals is assumed to be responsible for undertaking this work, and whether the work has been considered from an interprofessional perspective.

Collective Action
This is the operational work that people do to enact a set of practices, whether these represent a new technology or complex health care intervention. Typically, the implementation of new practices is seen as a management problem where the power lies to allocate resources and define the processes by which new technologies or complex interventions are executed into practice. In this construct, the key question is, How does the work get done? In particular, we are interested in work that defines and organizes the enacting of a complex intervention, in this case decision support for patients, into the routines of the existing workflows.

Reflexive Monitoring
This is the appraisal work by which employees assess and understand the ways that a new set of practices affects them and others around them. In this construct, the key question is, How is the work understood? Often operationalized as a set of measures or indicators, this construct is where data is collected about the work process (actions done) and work outcomes (results achieved). In the context of this book, monitoring is often viewed as counts of the number of patients deemed eligible for decision support, and the actual number who received, then viewed them.

---

*Please direct permission requests for Chapter 2 Section 2
to Glyn Elwyn at glynelwyn@gmail.com*

2.3 **Microsystems**
Marjorie M. Godfrey

CLINICAL MICROSYSTEM THEORY can trace its origins to many theoretical foundations including organizational development, improvement science, systems thinking, complexity science, psychology, medicine and health care, engineering, and business. One of the most influential pieces of work from business was James Brian Quinn's book *Intelligent Enterprise: A knowledge and service based paradigm for industry*[24]. Quinn, a professor from Dartmouth's Tuck School of Business, conducted international research to explore why some industry organizations, such as Merck, Honda, and Apple, were among the world's most dynamic and reputable businesses. Quinn introduced the idea of the SRU or smallest replicable unit as a venture where all frontline activity between employees and customers was intentionally designed to be its most efficient and to exceed customer expectations through optimized employee roles leading to high employee morale, high customer satisfaction, and exceptional business performance. In health care, the SRU is the frontline of care. It is this place where patients, families, and care teams meet that we call the clinical microsystem.

The microsystem is the place where (1) health care is delivered; (2) quality, value, safety, timely, reliability, efficiency, and innovation happens; and (3) staff morale and patient satisfaction are fostered. Microsystems include all professionals who regularly work together with a subpopulation of patients sharing a common purpose and information environment. The microsystem has processes, integrated technologies, and recurring patterns of information, behavior, clinical outcomes, practice performance, social interactions, and leadership[25].

From 2000 through June 2001, a team of Dartmouth researchers with funding from the Robert Wood Johnson Foundation studied a wide range of healthcare providers. The study, modeled after the Quinn service industry research, aimed to find the top-performing clinical microsystems in North America and to learn what made them different. What differentiates high-performing microsystems from others is their intentional development of five main characteristics: staff, leadership, patients, performance, information, and information technology (see Figure 2.3.1).

Microsystems form the building blocks of all health care organizations. The health care system quality is only as good as the quality produced by the smallest unit providing care, which is sometimes a lone health care worker

FIGURE 2.3.1
Success characteristics of high performing clinical microsystems

together with a patient. The following equation illustrates that the quality of health systems (QHS) is a sum of quality by microsystem (Q_m).

$$QHS = Q_{mn} + Q_{mn} + I$$

The true structure of the health system is composed of front line microsystems, and what we call "mesosystems" and the overarching "macrosystem." Surrounding these embedded systems is the geopolitical, regulatory, regional, state, and national context, since all systems live in a state, place, or region subject to different forms or types of government (see Figure 2.3.2).

In some cases patients receive care from one clinical microsystem, but more frequently the patient's path of care crosses multiple microsystems. When two or more microsystems are involved in the patient care journey, a mesosystem

FIGURE 2.3.2
Systems within systems in health care

Regional/Geopolitical	• State, National Context
Macrosystem	• Organizational or Facility
Mesosystem	• Two or more Microsystems
Clinical Microsystem	• Interprofessionals, Patients and Families Including Technology
Individual Care Provider	• Patient • Care Provider
Self-Care	• Patient • Family

is formed, although the mesosystem may not have the shared clinical and business aims, lined processes, and shared information environments that clinical microsystems strive to achieve (see Figure 2.3.3). Moving from studying only microsystems to studying the mesosystem, so as to learn about purpose, patients, professional processes, and patterns at this level, is essential improvement work to smooth the care path for patients and families. Nelson et. al. describe the health system's complexity as "an entire health care continuum" which is "easy to view" as "an elaborate network of microsystems that work together (more or less) to reduce the burden of illness for populations of people."[25]

FIGURE 2.3.3
Multiple microsystems create mesosystem

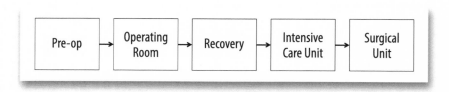

Clinical microsystems exist now, just as they always have since medicine was first practiced. Health care professionals can choose to study, improve, and lead them, or not. Activating clinical microsystems to build front-line capability to provide and improve care is a necessary skill for every health care professional to attain the desired health care outcomes and high performance. Absence of intelligent and dedicated improvement work by all staff in microsystems, mesosystems, and macrosystems, will prevent quality, value, efficiency, and pride in work from occurring and the burden of illness among populations will not be reduced.

Please direct permission requests for Chapter 2 Section 3 to Marjorie M. Godfrey at margiegodfrey@gmail.com

References

1. Gittell JH. Coordinating services across functional boundaries: The departure process at Southwest Airlines. In: Zemke R, Wood J, eds. *Best Practices in Customer Service: Case Studies and Strategies.* Amherst: HRD Press; 1998.

2. Gittell JH. Organizing work to support relational co-ordination. *The International Journal of Human Resource Management.* 2000;11(3):517–539.

3. Gittell JH. Supervisory span, relational coordination and flight departure performance: A reassessment of postbureaucracy theory. *Organization Science.* 2001;12(4):468–483.

4. Gittell JH. *The Southwest Airlines Way: Using the power of relationships to achieve high performance.* New York: McGraw-Hill; 2003.

5. Cramm JM, Nieboer AP. Relational coordination promotes quality of chronic care delivery in Dutch disease-management programs. *Health Care Management Review.* 2012;37(4):301–309.

6. Derrington T, Erikson-Warfield M. Engaging drug-exposed infants in early intervention services: The importance of relationships and communication with hospitals. *Topics in Early Childhood Special Education.* 2013.

7. Gittell JH. Coordinating mechanisms in care provider groups: Relational coordination as a mediator and input uncertainty as a moderator of performance effects. *Management Science.* 2002;48(11):1408–1426.

8. Gittell JH. *High performance healthcare: Using the power of relationships to achieve quality, efficiency and resilience.* New York: McGraw-Hill; 2009.

9. Gittell JH, Fairfield KM, Bierbaum B, Head W, Jackson R, Kelly M, Laskin R, Lipson S, Siliski J, Thornhill T, Zuckerman J. Impact of relational coordination on quality of care, postoperative pain and functioning, and length of stay: A nine-hospital study of surgical patients. *Medical Care.* 2000;38(8):807–819.

10. Gittell JH, Seidner R, Wimbush J. A relational model of how high-performance work systems work. *Organization Science.* 2010;21(2):490–506.

11. Gittell JH, Weiss L. Coordination networks within and across organizations: A multi-level framework. *Journal of Management Studies.* 2004;41(1):127–153.

12. Havens DS, Vasey J, Gittell JH, Lin W-T. Relational coordination among nurses and other providers: Impact on the quality of patient care. *Journal of Nursing Management.* 2010;18(8):926–937.

13. Gittell JH. Relationships and Resilience: Care Provider Responses to Pressures From Managed Care. *The Journal of Applied Behavioral Science.* 2008;44(1):25–47.

14. Gittell JH, Weinberg DB, Pfefferle S, Bishop C. Impact of relational coordination on job satisfaction and quality outcomes: A study of nursing homes. *Human Resource Management Journal*. 2008;18(2):154–170.

15. Havens DS, Vasey J, Gittell JH, Lin W-T. Impact of relational coordination on job satisfaction, emotional exhaustion and professional efficacy: A five hospital study of nursing. *Journal of Nursing Management*. 2012.

16. Warshawsky NE, Havens DS, Knafl G. The influence of interpersonal relationships on nurse managers' work engagement and proactive work behavior. *The Journal of Nursing Administration*. 2012;42(9):418–425.

17. Carmeli A, Gittell JH. High-quality relationships, psychological safety, and learning from failures in work organizations. *Journal of Organizational Behavior*. 2009;30(6):709–729.

18. Noël PH, Lanham HJ, Palmer RF, Leykum LK, Parchman ML. The importance of relational coordination and reciprocal learning for chronic illness care within primary care teams. *Health Care Management Review*. 2013;38(1):20–28.

19. Gittell JH. Relational coordination: Coordinating work through relationships of shared goals, shared knowledge and mutual respect. In: Kyriakidou O, Ozbilgin M, eds. *Relational Perspectives in Organizational Studies: A Research Companion*. Cheltenham: Edward Elgar Publishing Limited; 2006:74–94.

20. Gittell JH, Douglass A. Relational bureaucracy: Structuring reciprocal relationships into roles. *Academy of Management Review*. 2012;37(4):709–733.

21. Douglass A, Gittell JH. Transforming professionalism: Relational bureaucracy and parent-teacher partnerships in child care settings. *Journal of Early Childhood Research*. 2012;10(3):267–281.

22. Weinberg DB, Lusenhop RW, Gittell JH, Kautz CM. Coordination between formal providers and informal caregivers. *Health Care Management Review*. 2007;32(2):140–149.

23. May CR, Finch T. Implementing, embedding, and integrating practices: An outline of normalization process theory. *Sociology*. 2009;43(3):535–554.

24. Quinn JB. *Intelligent enterprise: A knowledge and service based paradigm for industry*. New York: The Free Press a division of Simon & Schuster Inc; 1992:48–63.

25. Nelson EC, Batalden PB, Godfrey MM. *Quality by Design: A Clinical Microsystems Approach*. 1st ed. San Francisco: Jossey-Bass, A Wiley Imprint; 2007.

SECTION TWO
CASE STUDIES AND COMMENTARIES

Many individuals who lead the work of implementing decision support for patients with the goal of achieving shared decision making presented their findings at the Summer Institute for Informed Choice in 2011 hosted by Dartmouth College in Hanover, New Hampshire. We wanted to capture the spirit of that Summer Institute, the enthusiasm and, just as often, the frustration, as we sensed the magnitude of the task ahead.

We also wanted to try and move the work towards the rapidly expanding field of improvement and implementation science, thereby situating the examples in wider frameworks to elicit factors that will facilitate, impede, or maybe even predict successful adoption of innovation into existing routines.

The following cases are followed by three commentaries. Jody Hoffer Gittell looks at each case from the perspective of relational coordination. Glyn Elwyn looks at the cases from the perspective of the Normalization Process Theory, based on the work of Carl May and colleagues. Marjorie M. Godfrey examines cases from the microsystems perspective. These three lenses guided debate at the Summer Institute and although many other implementation frameworks could have been used, we chose these three because they represent very different ways of exploring how best to put innovations into practice.

CHAPTER 3

Implementing Shared Decision Making in the United Kingdom:
Lessons from the Making Good Decisions in Collaboration (MAGIC) Program

Glyn Elwyn, Richard Thomson

Why we started: Informed by a report on the implementation of shared decision making in clinical settings, The Health Foundation in the United Kingdom requested proposals to test its practical implementation in the National Health Service (NHS). Research groups at Cardiff and Newcastle Universities collaborated with local health groups on this project.

What we set out to do: We wished to broaden the range of ways in which shared decision making could be considered by both the clinical team and the wider organization, and to test a range of methods for raising awareness and embedding shared decision making in routine practice.

What we achieved: By using Plan-Do-Study-Act (PDSA) cycles, the implementation work focused on both executives and clinical teams in the relevant organizations. For clinical teams, the work involved skills development in shared decision making and the use of quality improvement methods. At the organization level, we sought the support of executives, either at hospital board or, in primary care, at the level of partnerships. In secondary care, this was relatively easy at the board level but harder with middle managers and clinical directors. At the microsystem level of clinical teams it became clear that brief interventions (e.g., Option Grids and brief decision aids) and measures of decision quality were effective levers for change. The presence of clinical champions had significant impact on the teams' motivation.

What we learned: New methods are required to engage clinicians in this kind of implementation work, starting with revisions to clinical pathways to embed shared decision making that include skills development, multiple formats of patient decision support, and new methods of measurement that can provide rapid feedback.

CASE REPORT

Why we started: In 2010, the Health Foundation, an organization committed to improvement work in the UK health care sector, commissioned a report from Angela Coulter about the degree to which shared decision making had been implemented in clinical settings. Her report concluded that:

> Shared decision making attracts broad support from patients, professionals, and policy makers, yet it is infrequently implemented in clinical practice. Clinicians have been slow to respond to the evidence that a majority of patients, want to be involved in decisions about their care.[1]

Subsequently, the Health Foundation called for applications to test practical implementation of shared decision making in the NHS and to teach others wishing to introduce shared decision making into their clinical settings. With many trials of decision support interventions already published that indicated efficacy in controlled conditions, the Health Foundation decided to go beyond these narrowly focused studies[2] and test a range of different approaches, across different teams, to examine how best to change the culture of NHS organizations and make shared decision making part of clinical routines and workflows.

Research groups at Cardiff and Newcastle Universities had many years of experience in this field. The Cardiff group had published work on the competences of shared decision making,[3] developed measurement tools,[4] and developed and evaluated a number of web-based interventions[5,6]. The Newcastle group had examined the role of patient preferences[7] and had explored decision making in a range of clinical settings and contexts, including atrial fibrillation and stroke prevention, using decision analysis methods[8,9]. Both groups were aware of the need to decide how best to implement these interventions in routine settings, and the call enabled them to collaborate on a joint bid. This program, named "Making Good Decisions in Collaboration" (MAGIC), aimed to broaden the range of ways in which shared decision making could be considered at the clinical team and organization level. We proposed testing a range of methods for raising awareness and developing skills, hoping to learn how best to integrate shared decision making into NHS clinical routines in both primary and secondary care at Cardiff and Newcastle.

What we set out to do: We were commissioned to start the implementation program in August 2010, knowing already from several feasibility studies that implementation would not be easy[10]. Developments in improvement and implementation science over the last decade suggested that successful change required attention to wider organizational issues than efforts to implement shared decision making solely in team routines. Basing our plans on the premise that a multi-faceted approach was necessary, the MAGIC strategy placed a key focus on the clinical team but paid strong attention elsewhere, to dimensions that included influencing executives and boards, modifying, where possible, policy directives, introducing quality improvement initiatives, and using social marketing methods targeted at both professionals and patients. We developed a logic model to illustrate the influence drivers, mechanisms of influence, activities, effects and outcomes (see Figure 3.1).

In Cardiff, we recruited the following teams: ear, nose and throat, including the head and neck cancer team; breast cancer surgical units; and four primary care practices. In Newcastle, we recruited teams in urology, obstetrics, breast cancer surgery, and four primary care practices. We planned to engage the executive level (in the NHS, the Trust board or primary care partnership), to offer clinicians the opportunity to develop skills in shared decision making and facilitation, and to offer patient-targeted decision support tools. We also planned to use quality improvement methods to support rapid cycles of testing and feedback, using PDSA cycles. Where possible, we wanted to measure patient perceptions of their involvement (or lack of it) in clinical decisions and whether our efforts were having an impact by using qualitative methods (semi-structured interviews and field notes), as well as brief patient-reported surveys.

What we achieved: Our report described the progress made over the first 18 months of MAGIC, from August 2010 to March 2012. On the basis of this first phase, a further 18 months of work was commissioned by the Health Foundation to start in May 2012. The summary below is organized in broad headings that map to elements described in the logic model.

Organizational level policy and strategy: We found it relatively easy to achieve executive-level support for implementation work, at Trust and practice level. Board directors at both Cardiff and Newcastle took interest and ensured that reports were regularly made available to executive teams. It was much more difficult to ensure that this top-level commitment was espoused by other management tiers, particularly in secondary care. Middle managers and clinical directors who were not actively involved in the MAGIC clinical teams were often unaware of the program's aims and didn't understand their relevance to the wider organization. Furthermore, external factors, such as NHS standard metrics of interest, are not normally aligned to goals of shared decision

FIGURE 3.1
Logic model for the MAGIC program

#	Influence Drivers	Mechanisms	Activities	Effects
1	Policy, Strategic Leadership, Clinical Champions	Visible organisation strategic support.	Board level engagement, organisational mission, clinical leadership, set expectations and mandate measurement.	Microsystem pathway explicitly refer to SDM and the use of patient decision support.
2	Set Ethical Imperative			
3	**Professional Values:** patient-centered medicine, based on evidence-based practice	SDM principles espoused and expectations that teams use patient decision support.	Provision and skilled use of decision support, facilitation embedded in teams and engagement in skill development.	Teams who are able to 'do' and promote SDM, find and implement patient decision support tools.
4	Service and Quality Improvement Methods		Adopt agreed measures, collect, analyse and review data routinely, use PDSA cycles, facilitate clinical team development and pathway change.	Practice improvement culture, where SDM sets the standard of care.
5	Social Marketing	Develop suite of materials and activies to promote the SDM and use of patient decision support.	Raise awareness using branding, presentations, reminders and 'gifts,' engage using educational events and engagement with advocates and other stakeholders.	Professional motivation for providing shared decision making and a culture where supporting empowered informed patients are regarded as the norm.
6	Co-production Using Patient, Public and Professional Engagement	Engage stakeholders in the MAGIC programme.		

INFLUENCE DRIVERS → MECHANISMS → ACTIVITIES → EFFECTS

making: the Quality Outcome Framework in primary care,[11] for instance; or hospital waiting time targets that do not reward involving patients in treatment choices. On the other hand, when an alliance of middle management and clinical leadership did show interest and visited MAGIC program teams, we observed significant increases in motivation and activity.

Developing skills, knowledge, and attitudes at microsystem levels: When we first contacted clinical teams, we met varied responses. The clinical champions who had invited us to work with their teams in both primary and secondary care were supportive and curious to know how shared decision making would make a difference to their day-to-day work. But we also found many other clinicians who were either indifferent or reacted by saying "we already do shared decision making." This implied they were positive about the concept but also that they felt, contrary to empirical evidence and patient reports,[12] they had little to learn and no need to change. These initial reactions underwent a shift for most, though not all, clinicians over the next 6-9 months, facilitated by ongoing expert support, orientation to ideas of patient decision support, and skill development workshops. Short lunchtime events provided rapid orientation to principles of shared decision making and allowed opportunities for questions. Longer workshops used a three-step model of shared decision making[13] (see Figure 3.2) that emphasized the use of choice, option, and decision talk.

The MAGIC program also used prepared clinical scenarios involving simulated patients, so that clinicians could role-play shared decision making and begin using decision support tools. During these workshops we realized the value

FIGURE 3.2
Shared decision making: A model for clinical practice

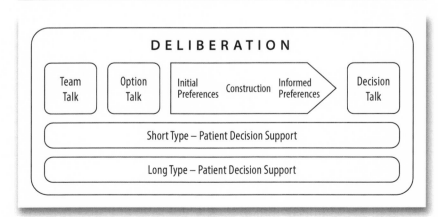

of using brief tools in clinical encounters, which led to a focus in Cardiff on Option Grids[14] and in Newcastle to the use of brief decision aids.

Quality Improvement methods: Quality improvement relied on identifying leadership, setting aims, and assessing progress through measurement for improvement and not measurement for judgment. It was difficult to introduce these ideas in parallel to shared decision making—teams, often skeptical to begin with or at least unfamiliar with them, easily conflated the two ideas, especially if they were new to both approaches. The use of PDSA cycles was particularly challenging because of the difficulties of measuring patients' experience with shared decision making. We made progress in two areas: (1) development and use of a short shared decision making questionnaire; and (2) a decision quality measure that included items to assess patients' knowledge and readiness to decide (DelibeRATE). Two teams in Cardiff and one in Newcastle used decision quality measures, providing evidence that they were clinically valuable because they helped practitioners assess how much their patients understood options, thus contributing to individual patient care as well as to monitoring of shared decision making. It became clear that embedding shared decision making and decisions tools and measures into practice would require revising clinical pathways—a goal we realize may not be sustainable over time; nor can we know whether these organizations will continue to value and make use of data generated by these tools.

Social Marketing efforts: Besides workshops directed at health care professionals, we also planned to inform patients and the wider public about shared decision making. We soon knew that we didn't have adequate time to develop a rigorous social marketing plan that involved careful collaboration with clearly identified target audiences, step-wise iterative consultation, and piloting. Still, besides our attention to clinicians we needed to influence patients. An Australian study had used three short questions to "activate" patients to become more involved in decisions,[15] and we modified some of their materials into posters, cards, and other reminders. Using rapid PDSA cycles, we tested the materials with a wide range of patients and professional groups, along with a short video for use in patient waiting areas. There were very positive reactions to the materials (see Figure 3.3) which we used in both Cardiff and Newcastle, although unfortunately without formal evaluation.

What we learned: Most, but not all clinicians, support shared decision making and readily agree that patients should be involved in decision making, an acceptance that is enhanced when clinicians are involved in skills development workshops. Providing tools—short ones as well as more

FIGURE 3.3
Ask 3 questions campaign at the Cardiff site of the MAGIC program

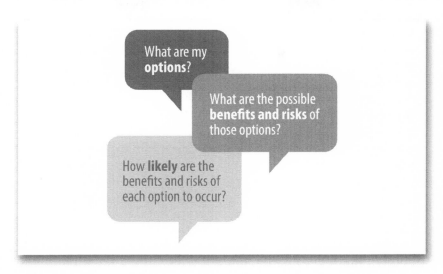

extensive tools like DVDs and websites—is helpful, but use depends on a positive attitude, skilled clinicians, and new clinical pathways that provide dedicated opportunities for patients and clinicians to use the tools effectively. For cultural change, we also found it extremely helpful to have support from the executive team and middle management, as well as "activating" patients with the expectation that they become participants in shared decision making. Measuring patient perception of engagement in shared decision making is difficult. Patients seem to interpret surveys as if they were being asked for their general satisfaction, and the results tend to be at ceilings with little variation. We do need better measures. There is no magic bullet when it comes to implementing shared decision making—it requires a multifaceted approach that addresses patients, health professionals, and health care teams as well as the organizational culture in which care is provided. We now have the benefit of further support from the Health Foundation to test our learning from MAGIC's Phase 1 in a second, 18-month, Phase 2. For more information about the MAGIC program, see the following websites: www.health.org.uk/areas-of-work/programmes/shared-decision-making and shareddecisionmaking.health.org.uk.

Acknowledgements

The authors wish to acknowledge the work done by the following collaborators in the MAGIC Program. In Cardiff: Natalie Joseph-Williams, Andrew Rix, Amy Lloyd, Emma Cording, Mike Spencer, Helen Mcgarrigle, Annette Beasley, Val Willmott, Adrian Edwards, Jan Davies, Alun Tomkinson, Keith Cass, Laura Roach. In Newcastle: Richard Thomson, Carole Dodd, Lynne Stobbart, Dave Tomson, Diane Palmer, Sheila Macphail, Steve Robson, Claire Leader, Chris Watson, Rob Pickard, Jill Ferguson, Chris Hall, Matt Shaw, Stewart Nicholson, Sandra Scott, Hume Hargreave, Julie Marshall, Jo Lally.

COMMENTARIES

Relational Coordination Theory: Jody Hoffer Gittell

The MAGIC program's major innovation was Option Grids and brief decision aids, examples of in-encounter tools for facilitating shared decision making between patients and care providers. Option Grids and brief decision aids help support relational coproduction between the patient, family, and care providers; and also relational coordination among the care providers themselves, keeping everyone on the same page through greater shared knowledge. MAGIC's second innovation was changing clinical pathways to support relational coproduction and relational coordination, thereby creating clarity on who is doing what and when, while strengthening shared knowledge and shared goals. The program's third innovation was new methods of measurement, not only regarding use of decision making tools, as other projects in this book have done, but measuring patient perceptions of shared decision making itself. Performance measures that focus attention on outcomes requiring every participant's cooperation help create shared goals. MAGIC's fourth innovation was investing in the development of frontline care provider skills.

The Relational Coordination Theory argues that three kinds of interventions—*structural, relational,* and *work process*—are needed to strengthen relational coordination and coproduction. The MAGIC program's innovations as reported here have been primarily *structural interventions*: Option Grids and brief decision aids, revised clinical pathways, new forms of performance measurement, and new training methods to support new patterns of interaction. But changing any clinical pathway requires a *work process intervention* to reorganize the flow of tasks. Staff engagement in these structural and work process interventions opens up opportunities for *relational interventions*, such as relational mapping and interdependence conversations, with potential to deepen MAGIC's impact. These aspects of the change process are less well described and provide good opportunities for further work.

Normalization Process Theory: Glyn Elwyn

By examining the logic model in the MAGIC case we trace how the implementation efforts could contribute to a normalization process. Participating teams had to be clear about the new work; short introductory sessions and training workshops helped with this. Coherence, or making sense of the new work toward accomplishing shared decision making, was important; the unfolding narrative here suggests, however, that coherence was only partly achieved among the relevant clinical teams. Among team members, there was disagreement and confusion as to what constituted good decision making processes in health care, and so communal specification was not successfully achieved, therefore limiting internalization for many critical team members. The other three constructs—cognitive participation (agreement about who does the work), collective action (how the work gets done), and reflexive monitoring (how the work is measured)—rely heavily on successfully achieving coherence. Execution of innovations such as Option Grids, plus partial implementation of the Ask 3 Questions approach, as well as degrees of measurement (using Decision Quality Measures) newly achieved—all speak to the MAGIC team's ability to pull together excellent working relationships with clinical teams in the relevant NHS organizations. However, the MAGIC team would be first to admit that shared decision making was not fully embedded into routine practice, with normalization never fully achieved.

Microsystems Approach: Marjorie M. Godfrey

The MAGIC program's strategy included engagement at multiple levels of the health care system. Recalling the "Systems within Systems" diagram illustrating the "embeddedness" of multilevel systems, MAGIC sought and obtained executive leadership support including at the board or macrosystem level. At the geopolitical level (this would be the NHS itself), the organization didn't align with the shared decision making program and didn't "incentivize" this process by rewarding patient involvement in treatment choices, therefore missing an opportunity to further strengthen support at all levels of the health system. Mid-level managers and clinical directors in the mesosystem weren't adequately informed of the program and subsequently couldn't advocate to those with authority to allocate resources. This is illustrated very clearly in the MAGIC case and represents typical mesosystem dysfunction in many organizations.

A disconnect between the macrosystem and the microsystem manifests itself in the mesosystem where support, leadership, and expectations make the difference for an organization's execution strategy. What is encouraging, in this particular case, is the actions of mid-level managers or mesosystem leaders who were interested and actively engaged with the microsystem teams. The resulting increase in motivation and activity in just this one program exemplifies the mesosystem's power to influence the microsystem. High performing microsystems emphasize staff development. Continued facilitation by expert support encourages learning and skills development, within the microsystem, in several ways. Learning about shared decision making is an excellent means of improving care. Including the patient and focusing on clinical pathways further illustrates the clinical microsystems' role in acquiring shared decision making skills and integrating them into the pathway of care.

Reflection on Commentaries
The discussions at the Summer Institute in 2011 applied differing implementation frameworks to critique progress within phase 1 of MAGIC, throwing new light on this and other implementation programs. Reflecting back from a position halfway through phase 2 of MAGIC, which is designed to test learning from phase 1, these perspectives still resonate.

In light of subsequent learning, I would like to reiterate the importance of a multifaceted approach towards implementation of these programs, but would also emphasize the critical role of skills training as a key change agent for clinical attitudes and practice. Nonetheless, implementation remains challenging and success at a local level is often dependent upon wider system-level barriers and incentives. Furthermore, measurement from the patient perspective remains highly challenging.

Richard Thomson

Please direct permission requests for Chapter 3 to Glyn Elwyn at glynelwyn@gmail.com

References

1. Coulter A. *Implementing shared decision making in the UK: A report for the health foundation.* London; 2009:50.

2. Stacey D, Bennett CL, Barry MJ, Col NF, Eden KB, Holmes-Rovner M, Llewellyn-Thomas HA, Lyddiatt A, Légaré F, Thomson RG. Decision aids for people facing health treatment or screening decisions. *Cochrane Database of Systematic Reviews.* 2011;10(10).

3. Elwyn G, Edwards AG, Kinnersley P, Grol R. Shared decision making and the concept of equipoise: The competences of involving patients in healthcare choices. *The British Journal of General Practice.* 2000;50(460):892–899.

4. Elwyn G, Hutchings H, Edwards AG, Rapport F, Wensing M, Cheung W, Grol R. The OPTION scale: Measuring the extent that clinicians involve patients in decision-making tasks. *Health Expectations.* 2005;8(1):34–42.

5. Evans R, Elwyn G, Edwards AG, Newcombe R, Kinnersley P, Wright P, Griffiths J, Austoker J, Grol R. A randomised controlled trial of the effects of a web-based PSA decision aid, Prosdex. Protocol. *BMC Family Practice.* 2007;8:58.

6. Sivell S, Marsh W, Edwards AG, Manstead ASR, Clements AM, Elwyn G. Theory-based design and field-testing of an intervention to support women choosing surgery for breast cancer: BresDex. *Patient Education and Counseling.* 2012;86(2):179–188.

7. Say R, Murtagh M, Thomson RG. Patients' preference for involvement in medical decision making: A narrative review. *Patient Education and Counseling.* 2006;60(2):102–114.

8. Robinson A, Thomson RG. The potential use of decision analysis to support shared decision making in the face of uncertainty: The example of atrial fibrillation and warfarin anticoagulation. *Quality in Health Care.* 2000;9(4):238–244.

9. Thomson RG, Parkin D, Eccles M, Sudlow M, Robinson A. Decision analysis and guidelines for anticoagulant therapy to prevent stroke in patients with atrial fibrillation. *Lancet.* 2000;355(9208):956–962.

10. Elwyn G, Scholl I, Tietbohl C, Mann M, Edwards AG, Clay C, Légaré F, Van der Weijden T, Lewis CL, Wexler RM, Frosch DL. Many miles to go...: A systematic review of the implementation of patient decision support interventions into routine clinical practice. *BMC Medical Informatics and Decision Making.* 2013;In Press.

11. Ashworth M, Armstrong D. The relationship between general practice characteristics and quality of care: A national survey of quality indicators used in the UK Quality and Outcomes Framework, 2004-5. *BMC Family Practice.* 2006;7:68.

12. Coulter A. *Engaging Patients in Healthcare.* New York: Open University Press; 2011.

13. Elwyn G, Frosch DL, Thomson RG, Joseph-Williams N, Lloyd A, Kinnersley P, Cording E, Tomson D, Dodd C, Rollnick S, Edwards AG, Barry MJ. Shared decision making: A model for clinical practice. *Journal of General Internal Medicine.* 2012;27(10):1361–1367.

14. Elwyn G, Lloyd A, Joseph-Williams N, Cording E, Thomson RG, Durand M-A, Edwards AG. Option Grids: Shared decision making made easier. *Patient Education and Counseling.* 2013;90(2):207–212.

15. Shepherd HL, Barratt A, Trevena LJ, McGeechan K, Carey K, Epstein RM, Butow PN, Del Mar CB, Entwistle VA, Tattersall MHN. Three questions that patients can ask to improve the quality of information physicians give about treatment options: A cross-over trial. *Patient Education and Counseling.* 2011;84(3):379–385.

CHAPTER 4

Implementing Decision Support and Shared Decision Making:

Steps Toward Culture Change

Suepattra G. May, Caroline Tietbohl, Dominick L. Frosch

Why we started: This case study describes a demonstration project at a large multi-specialty physicians' group practice in Palo Alto, California that aimed to address gaps in the real-world implementation of decision support interventions (DESIs) while building a sustainable model of shared decision making in primary care.

What we set out to do: Our approach entailed: (1) working collaboratively with clinic stakeholders to tailor DESI implementation to site-specific workflows; (2) establishing our program within the organization by means of a social marketing campaign; and (3) frequently engaging clinic staff and physicians through academic detailing visits.

What we achieved: 121 clinic staff, 58 doctors, and one nurse practitioner distributed DESIs for 16 different topics, for a total of 4,892 DESIs distributed across all participating clinics. The percentage of potentially eligible patients who received a DESI ranged from 0.2% to 44.6% for colorectal cancer screening, and from 0.0% to 32.6% for back pain.

What we learned: We identified clinic staff involvement beyond the physician and patient activation as facilitators for raising awareness of shared decision making where attitudes of both patients and physicians towards shared decision making could potentially serve as barriers. Adopting shared decision making in clinical practice will require sustained effort toward genuine cultural change.

CASE REPORT

Why we started: Widespread adoption of decision support interventions (DESIs) to promote shared decision making has not occurred in primary care practice, despite their proven efficacy in improving patients' knowledge and expectations, and in aligning patients' values with treatment choices, for a variety of conditions[1]. Furthermore, we lack information regarding implementation of DESIs into primary care in a multi-specialty fee-for-service setting.

This case study describes the efforts of a demonstration project at the Palo Alto Medical Foundation for Healthcare, Research and Education (PAMF), a 501(c)(3) not-for-profit medical foundation serving over 700,000 patients in the San Francisco Bay Area. PAMF contracts with various payers, owns and operates facilities including two ambulatory surgical centers, and contracts with an independent medical group, the Palo Alto Foundation Medical Group (PAFMG) for medical services. The group currently has over 100 physicians and 4300 employees working in both primary and specialty care, and practicing in 40 PAMF clinics. Six of these clinics, including five primary care clinics, one specialty care clinic, and three education resource centers, participated in the demonstration project.

In partnership with the University of California, San Francisco, the investigative team sought to address important gaps in real-world DESI implementation while building a sustainable model of shared decision making and DESI use in primary care. While both PAMF leadership and clinicians believed that integrating shared decision making and DESIs into primary care practice was an integral part of achieving efficient, effective, and patient-centered care, DESIs have not prominently or consistently been used in practice. Significant challenges remain to creating a sustainable model, among them determining efficient ways to identify patients at appropriate times to engage in shared decision making, incorporating DESIs into clinic workflows, determining effective delivery methods, and assessing the effects of DESIs on resource utilization[2].

What we set out to do: We recognized early on how incorporating patient decision support and adopting shared decision making in routine clinical practice required a culture change among healthcare providers. Although PAMF prides itself on being patient-centered, changing practice workflows and introducing new tools requires sustained efforts to ensure that changes become routine and lasting. For these reasons, we developed a multi-pronged social marketing strategy in the participating demonstration project clinics[3]. We aimed to work collaboratively with clinic stakeholders to understand their individual clinic cultures and to tailor initiatives to their site-specific workflows.

Clinic stakeholders selected which DESIs would be most suitable in their respective clinics and what would be the best modes of delivery (e.g., physician prescription, medical assistant warm hand-off, patient request). We hoped to generate robust quantitative and qualitative data about what happens when DESIs are made routinely available in community-based practices by:

1. Raising awareness of and establishing our program within the organization by means of a social marketing campaign with its own program name;
2. Frequently engaging clinic staff and physicians through academic detailing visits;
3. Collecting detailed ethnographic field notes to document the implementation process;
4. Analyzing data comparing health care utilization among patients from clinics participating in the demonstration project with comparable non-participating clinics or clinics that did not have access to the same DESIs.

Our social marketing strategy, meant to give our program and message repeated exposure, incorporated techniques used by the pharmaceutical industry to market new products[4]. In addition to creating our own program identity—the *Partners in Medical Decision Making Program* (PMDM)—we created program branding using two taglines that conveyed its purpose to patients, staff, and physicians: (1) *Prescription Strength Information for Better Decisions*; and (2) *Better Decisions, Together*. We developed publicity materials, including posters and brochures, which we displayed and distributed in clinic lobbies, waiting areas, and exam rooms to promote the program directly to patients. We then designed PMDM-branded promotional items, such as pens, Post-it notes, mints, water bottles, tote bags, pedometers, lunch bags, and mugs, chosen in collaboration with our clinic stakeholders as useful items that would create the desired repeated exposure to the PMDM messages and brand. Samples of these branded materials can be seen in Figure 4.1.

What we achieved: Since the program's inception in January 2010, we distributed 4,892 DESIs across all clinic sites. The participating sites have distributed a total of 16 different decision aid topics, with colorectal cancer screening and advanced directives being distributed most frequently (totals of 2,020 and 807, respectively). Across all participating sites, 121 clinic staff, 58 physicians, and one nurse practitioner distributed decision aids to patients. Of the patients who received decision aids, 41.7% were men and 58.3% were women, with patient ages ranging from 10 to 98, the average patient age being 60.

FIGURE 4.1
Examples of social marketing campaign items with PMDM brand name and tag lines

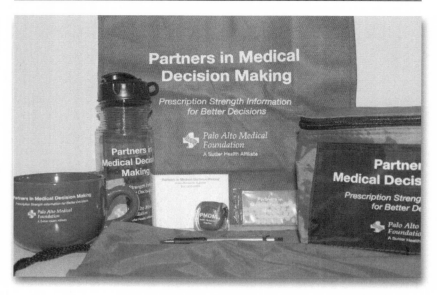

Continuous provider engagement was a major component of our program strategy. Early on, clinic staff and physicians expressed interest in receiving feedback about their distribution numbers and the number of eligible patients receiving DESIs. To foster competitive spirit within and among clinics, we worked with PAMF data management personnel to identify denominator data for two DESI health topics: colorectal cancer screening and back pain. Across the four clinics distributing colorectal cancer screening DESIs, the percentage of potentially eligible patients who received a DESI ranged from 0.2% to 44.6%. Across the three clinics distributing for acute and chronic back pain, the average percentage of potentially eligible patients who received a back pain DESI ranged from 0.0% to 32.6% (see Figure 4.2).

We also employed several measures to elicit patient engagement. First, the posters and brochures we created to advertise our program allowed patients to request DESIs independently of their health care team. Second, at one clinic we piloted a patient self-screener to facilitate patient activation for the colorectal cancer screening DESI. The screener was distributed upon check-in to patients age 50 and older. Patients responded to questions about eligibility for colorectal cancer screening and were instructed to present the screener to staff if they were interested in receiving a DESI. During the month that the

FIGURE 4.2
Percent of eligible patients who received DESIs about colorectal cancer screening and back pain[3]

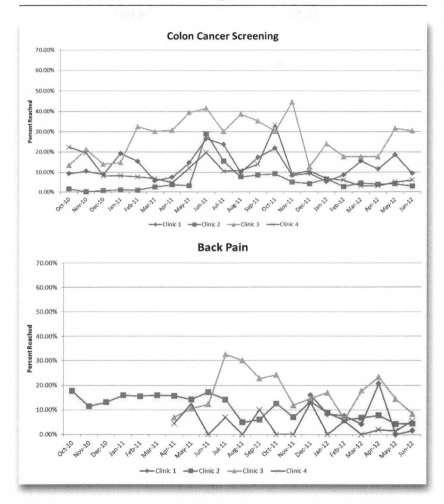

screener was implemented, colorectal cancer screening DESI intervention distribution was five times greater than during the previous month.

In addition to regular feedback reports, we provided PMDM promotional items as "recognition and rewards" for clinic staff as an attempt to maintain program awareness. For example, we gift-wrapped pens, Post-its, and a PMDM information card as a "stationery kit" gift, and put various items within a PMDM tote as a "goody bag." We also strived to keep pivotal PAMF stakeholders informed and aware of our program efforts. Twice a year we delivered a PMDM Executive Report distributed in hard copy newsletter format to PAMF leadership. With each report we provided promotional materials such as mugs, water bottles, tote bags, pens, and Post-it notes to serve as continuous reminders of our program even after the report had been read.

What we learned: During our three-year implementation program, we learned that (a) clinic staff involvement beyond the physician and (b) patient activation may help raise awareness of shared decision making in community settings. By working directly with clinics to design workflows tailored to their individual cultures, we fostered enough initial interest in distributing DESIs to create processes that integrated into existing workflows. Regular academic detailing by the project team was also crucial for the launch and maintenance of DESI distribution. Inviting patients to engage in shared decision making through posters in exam rooms, brochures in waiting areas, or through patient self-screeners appeared to facilitate patient engagement. However, many barriers to the provision of regular decision support to patients still remain. Physicians frequently cite time as a key barrier, while reporting that they believe they already engage in shared decision making.

Our work on both this and other projects has identified additional barriers entrenched in medical culture. Many patients feel vulnerable when asserting their preferences with physicians and playing an active role in clinical decision making, fearing that they may be viewed as "difficult" patients and that this could lead to worse care[5]. We have begun chipping away at these barriers, but structural incentives signaling to health care providers that shared decision making and decision support is expected of their care, are necessary, in addition to helping patients understand that there are often options and their preferences matter when making clinical decisions. Implementation of shared decision making and decision support thus requires a major cultural shift from both physicians and patients[1]. While many patients have embraced this shift, many physicians still resist substantive changes in their practice.

Acknowledgements
Our work was supported by a grant from the Informed Medical Decisions Foundation (IMDF).

COMMENTARIES

Relational Coordination Theory: Jody Hoffer Gittell

The Palo Alto Medical Foundation effort stands out for its (1) deliberate inclusion of the entire clinical staff beyond the physician, (2) tailoring distribution of decision support interventions (DESIs) to each clinic's unique workflow, and (3) use of social marketing to motivate changes in staff behaviors.

The change effort increased DESI distribution, although we don't know whether an actual increase in shared decision making, beyond distribution of the tools, also occurred. I would surmise that by engaging the whole clinical staff in distribution, there may have been an increase in the overall strength of shared goals, shared knowledge, and mutual respect among staff, as well as more frequent, timely, accurate, and problem-solving communication; and that these dynamics (known as relational coordination) likely helped foster similar dynamics between staff and patients (known as relational coproduction), thereby increasing staff's ability to engage patients and families in shared decision making.

While both *relational* and *work process interventions* were used in this inclusive change process to encourage improvements in both relational coordination and co-production, basic structures were not altered to sustain these changes. May, Tietbohl, and Frosch observed continued barriers to shared decision making from patient concerns about speaking up to physicians' beliefs that they already engage in shared decision making. They recommend structural interventions, such as changes in reward structures, to address these barriers. I would concur, once there is evidence that relational dynamics have indeed begun changing, and that physicians are part of these improved dynamics. Though they cannot take the place of relational interventions for changing underlying relational dynamics toward stronger shared goals, shared knowledge, and mutual respect between physicians and their patients, structural interventions like rewards can help sustain changes already in progress.

Normalization Process Theory: Glyn Elwyn

The Palo Alto case outlines partial success (at least for distribution of tools), and in many aspects mirrors the narrative of other cases in this volume. The authors admit that despite their considerable efforts in advocating use of patient DESIs and using social marketing techniques coupled with regular "detailing" visits, the numbers of eligible patients who were provided with the tools were relatively low; the impression is that of an uphill struggle to sustain interest in distribution of these tools. When examined through the NPT lens the challenge becomes more evident. The goal was to affect change across a very large organization, across many units, where there was resistance to the proposed change in practice (as the researchers admit), particularly among the clinicians that they hoped would provide patients with the tools. Despite efforts to inform the clinical teams about tools and their benefits, there does not seem to be, from an NPT standpoint, any evidence of coherence. The tools are only a small part of the work of involving patients in decisions; other core components include attitudes, motivation, and skills to undertake the face-to-face work with patients that constitute accomplishment of the proposed new work. NPT asks whether participants enjoyed consensus on who does the work (cognitive participation), how the work gets done (collective action), and how is the work measured (reflexive monitoring). Although we hear of measures (e.g., numbers of tools distributed), this case, like most others, indicates that there was not much progress towards these four critical constructs of Normalization.

Microsystems: Marjorie M. Godfrey

The PAMF case stresses the importance of each clinical practice's unique culture and how any improvements or changes must recognize its unique characteristics. This is a basic underpinning of the clinical microsystem; each has its own culture, its own patient populations, professionals, processes, and patterns. Allowing individual clinical microsystems authority over the distribution, marketing, and sharing of interventions with patients encouraged an increase in adoption, despite variable distribution rates. Additionally, the group recognized the daily work of the interprofessionals in the practices and identified where shared decision making could be integrated into workflow.

The group seemed to focus solely on physician contribution while obscuring (for readers) which clinic staff were involved. The opportunity existed to ensure that everyone in the practice was engaged in identifying possible improvements and redesign of daily workflow. The marketing campaign provided constant communication to patients and professionals about shared decision making, which resulted in more patient-activated discussions. The report cited physicians' time as a barrier to shared decision making. This forces one to wonder how, if other members of the practice were involved, might their participation lessen this burden reported by physicians.

The macrosystem offered incentives, seemingly an expectation of care delivery, to promote shared decision making and decision support. The case concludes that shared decision making and decision support requires a major cultural shift by physicians and patients. But cultural transformation is essential for all members of the microsystem. Including patients and families in assessing current workflows and redesigns within a shared decision making process would result in a stronger sense of ownership by all contributors to the microsystem.

Reflection on Commentaries

We concur with the editors' assessments. Despite our intensive social marketing and academic detailing strategies, our experience highlights the structural and cultural barriers in health care delivery that impede more widespread and systematic distribution of DESIs in clinical practice. Future DESI implementation projects should incorporate one or more of the three theoretical approaches from project inception. Allowing for real time analysis, evaluation, and reconfiguration of a DESI implementation strategy through the relational coordination, NPT, or microsystems lenses makes possible a more comprehensive understanding of what is needed to facilitate culture change, increase receptivity towards shared decision making, and encourage uptake of DESIs.

Suepattra G. May, Caroline Tietbohl, Dominick L. Frosch

Please direct permission requests for Chapter 4 to Suepattra G. May at mays@pamfri.org

References

1. Stiggelbout AM, Van der Weijden T, De Wit MPT, Frosch DL, Légaré F, Montori VM, Trevena LJ, Elwyn G. Shared decision making: Really putting patients at the centre of healthcare. *BMJ*. 2012;344(1):e256.

2. Frosch DL, Moulton BW, Wexler RM, Holmes-Rovner M, Volk RJ, Levin CA. Shared decision making in the United States: Policy and implementation activity on multiple fronts. *Zeitschrift für Evidenz, Fortbildung und Qualität im Gesundheitswesen*. 2011;105(4):305–312.

3. Luck J, Hagigi F, Parker LE, Yano EM, Rubenstein LV, Kirchner JE. A social marketing approach to implementing evidence-based practice in VHA QUERI: The TIDES depression collaborative care model. *Implementation Science*. 2009;4(1):64.

4. Fugh-Berman A, Ahari S. Following the script: How drug reps make friends and influence doctors. *PLoS Medicine*. 2007;4(4):e150.

5. Frosch DL, May SG, Rendle KAS, Tietbohl C, Elwyn G. Authoritarian physicians and patients' fear of being labeled "difficult" among key obstacles to shared decision making. *Health Affairs*. 2012;31(5):1030–1038.

CHAPTER 5

Shared Decision Making at Group Health:
Distributing Patient Decision Support and Engaging Providers

Clarissa Hsu, David T. Liss, Emily O. Westbrook, David Arterburn

Why we started: Driven by a combination of strong organizational leadership and Washington State's political context, Group Health, a consumer-governed, non-profit health system, committed itself to integrating patient decision aids and shared decision making into routine practice. Although randomized controlled trials show that patient decision aids can improve decision quality, integration of decision aids into routine practice has been slow and so we aimed to study the process of implementation.

What we set out to do: In January 2009, Group Health initiated a large-scale implementation of video-based patient decision aids for 12 preference-sensitive health conditions related to elective surgical procedures. The primary goal was to incorporate decision aids into standard clinical practice for six specialty service lines within Group Health's Western Washington Group Practice Division, serving approximately 366,000 members.

What we achieved: Over two years, approximately 10,000 decision aids were distributed through system-level and clinic-specific processes developed to integrate 12 video-based decision aids into standard practice. This was achieved through strong leadership, financial support, a well-defined implementation strategy, and a commitment to engaging in process improvement throughout implementation.

What we learned: We identified several factors as important for successful implementation. Strong support from senior leaders as well as a system for pre-visit ordering helped with sustainability and follow-up. Providing timely feedback to teams about distribution rates while engaging providers and staff in implementation process development secured group buy-in. We still need to find ways to address concerns about conditions perceived as life-threatening and/or time-sensitive.

CASE REPORT

Why we started: Group Health is a Washington State consumer-governed health care organization committed to providing affordable health care and specifically emphasizing the value and importance of preventive care. Group Health's integration of health care delivery with health care insurance allows it to minimize fee-for-service medicine. Salaried physicians in the group practice are encouraged to provide care based on evidence-based guidelines and patient-centered decision making. Despite existing structures that should have minimized variation, there were, however, considerable differences in elective surgery rates—including knee and hip replacement, low back surgery, and hysterectomy—during the years 1994-2007.

In 2007, Washington State passed legislation to promote shared decision making demonstration projects, identifying shared decision making with patient decision aids as a higher standard of informed consent[1]. Randomized controlled trials show that decision aids improve patient knowledge, patient satisfaction with the decision making process, and concordance between patient values and treatment choices[2-6]. Often, the use of decision aids for preference-sensitive decisions results in patients choosing less invasive treatments[3]. Nonetheless, integration of patient decision aids into everyday clinical practice has been slow[7].

Senior clinical leaders in Group Health endorsed implementation of a shared decision making demonstration project and committed significant organizational resources toward supporting the work, emphasizing the importance of improving clinical outcomes and controlling costs by reducing unwarranted variations in care. Specialty care leaders mandated the use of decision aids by frontline providers and staff, maintaining that positive outcomes from decision aids would include better informed patients, improved quality of care, and greater liability protection under the new legislation. Although leaders hoped that the rates of some procedures and care costs might decrease with the use of decision aids for elective surgery, this was not the main message given to frontline providers. Because leaders elected to first deploy decision aids in specialty service lines before using decision aids in primary care, our evaluation efforts focused on implementation and provider perceptions in the six participating specialty care service lines.

What we set out to do: This project began in January 2009 with large-scale implementation of video-based patient decision aids at Group Health for 12 preference-sensitive health conditions related to elective surgical procedures (see Table 5.1). The primary aim was incorporating decision aids into standard

TABLE 5.1
Summary of decision aids, treatment choices, and total distribution

SERVICE LINE, TOPIC (length of decision aid video)	TREATMENTS	DISTRIBUTIONS as of January 2011
ORTHOPEDICS		
Hip osteoarthritis (45 minutes)	Surgery, medical therapy	1310
Knee osteoarthritis (54 minutes)	Surgery, medical therapy	3469
CARDIOLOGY		
Coronary artery disease (51 minutes)	Surgery, percutaneous coronary intervention (PCI) medical therapy	706
WOMEN'S HEALTH		
Uterine fibroids (54 minutes)	Surgery, medical therapy	956
Abnormal uterine bleeding (32 minutes)	Surgery, medical therapy	995
UROLOGY		
Benign prostatic hyperplasia (39 minutes)	Surgery, medical therapy	880
Prostate cancer	Surgery, radiation, hormone therapy, active surveillance	347
GENERAL SURGERY		
Early stage breast cancer surgery (54 minutes)	Lumpectomy, mastectomy, radiation, hormone therapy	168
Breast reconstruction (55 minutes)	Reconstructive surgery or no reconstruction	12
Ductal carcinoma in situ (54 minutes)	Lumpectomy, mastectomy, radiation, hormone therapy	49
NEUROSURGERY		
Low back pain: spinal stenosis (36 minutes)	Surgery, medical therapy	520
Low back pain: herniated disc (39 minutes)	Surgery, medical therapy	415
TOTAL DECISION AIDS DISTRIBUTED:		**9827**

clinical practice for six specialty service lines within Group Health's Western Washington Group Practice Division, which serves approximately 366,000 members. The Informed Medical Decisions Foundation (IMDF) and Health Dialog were the partners for this demonstration project, with Health Dialog providing free decision aids for approximately two years, after which a contract was negotiated for their purchase. The Group Health Research Institute received funding from the Group Health Foundation and the Commonwealth Fund to evaluate the demonstration project (for a more detailed case study see our upcoming article in *Medical Decision Making* [8]). While other decision aid formats were considered, Group Health selected Health Dialog's decision aids because they were considered the most comprehensive and evidence-based.

The next step was working sequentially with each specialty service line to develop decision aid distribution systems. Two senior Group Health project management consultants, at approximately 1.15 FTE for two years, planned and guided implementation tailored to the unique patient needs and workflow of each specialty service line and clinical location. Consultants and service line leaders drafted distribution processes that were shared with frontline providers and staff. Provider and staff reactions and input helped revise the distribution process, usually in one meeting, but in multiple meetings if necessary. Negotiation periods ranged from six weeks to three or four months. After an implementation process was established, a "go-live" date was set, with project managers visiting each clinic site at least once to monitor implementation processes and progress. Additional visits and calls to sites were conducted as necessary.

The video-based decision aids were primarily distributed as physical DVDs and booklets. Decision aids were ordered for patients by clinical staff through Group Health's electronic health record (EHR, EpicCare, Epic Systems Corporation). An existing Group Health service that sends educational materials to patients via US mail received and executed orders. The DVDs were also viewable online via the MyGroupHealth patient portal. Providers could include a link in the patient's after-visit summary. Although decision aids could be distributed in-office, the consultants had trouble tracking decision aids distributed in the clinic quite early in the project. Since tracking distribution rates was critical for evaluation, in-office distribution was strongly discouraged, with exceptions for Cardiology and Breast Cancer treatment options. The time between an initial appointment and a procedure was too short for the mailed option.

To facilitate productive conversations between patients and providers, the goal was to deliver the decision aid before the patient's first specialty consult.

FIGURE 5.1
Monthly decision aid distribution by service line

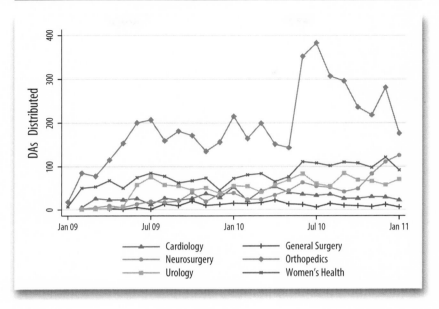

However, specialty providers reported reluctance to implement a pre-visit decision aid delivery for some health conditions because of concerns about the accuracy of the referral diagnosis. For example, Women's Health providers were concerned about distributing decision aids to women referred for abnormal uterine bleeding before conducting a thorough evaluation for possible cancerous causes of bleeding. In these cases, the provider was responsible for ordering the decision aid at the point of care and ensuring that a follow-up conversation occurred after the patient had viewed the decision aid.

What we achieved: By January 2011, two years after the start of the demonstration project, Group Health had established system-level and clinic-specific processes that facilitated the distribution of decision aids. Over two years, approximately 10,000 decision aids were distributed (see Figure 5.1).

Decision aid distribution began in the Orthopedic and Women's Health service lines in January 2009, followed by Cardiology and Urology. General Surgery began distributing decision aids in spring 2009, coordinating with Group Health programs that distributed educational materials to women diagnosed with breast cancer. Separate Neurosurgery distribution protocols were created in March 2010 to accommodate administrative challenges and leadership changes.

Sixteen months into the project implementation a process improvement effort occurred in connection with Group Health's adoption of Lean management principles[9-12]. The result was major revisions to decision aid workflows in several service lines. For example, patient visit schedulers received a list of words and phrases a patient might mention while scheduling an appointment, any of which could identify the patient as a decision aid candidate. As a result, schedulers were instructed to notify the specialty service provider's team by email if these words were mentioned. A cover letter added to the mailed decision aids emphasized that the decision aid did not replace a specialist consult and that not all treatment options in the decision aid were appropriate for all patients. These changes coincided with increased decision aid distribution in some service lines, particularly for conditions that were easily identified before a specialist visit (e.g., knee and hip osteoarthritis) (see Figure 5.1).

Distribution gradually decreased, however, consistent with most quality improvement initiatives. Finally, in fall 2010, a modification in the electronic medical record referral process resulted in a new prompt that asked primary care providers if they wanted to order the decision aid when referring to a specialist for the selected conditions. This resulted in a strong shift towards primary care providers ordering decision aids, although emphasis remained on specialists having shared decision making conversations with patients; and primary care providers were not given additional training in shared decision making.

What we learned: Our qualitative data collection identified several lessons learned:

- Strong support from senior leaders was critical to launching and sustaining the project;
- Pre-visit ordering and timely feedback to teams about distribution rates helped facilitate distribution and follow-up;
- Including providers and staff in devising the implementation process increased buy-in;
- Strategies to address concerns about conditions perceived as life-threatening and/or time-sensitive were needed.

Additional facilitators and barriers to implementation appear in Table 5.2. Below we provide details about a few critical lessons learned in order to aid those interested in implementing a similar process in a large health care organization.

TABLE 5.2
Facilitators and barriers to effective decision aid implementation – All service lines

FACILITATORS	
Workflow and Infrastructure	• High-quality decision aid at little or no cost • Pre-existing mechanisms for distributing condition-specific materials • Ability to add pre-visit decision aid ordering into workflow • Letter to patients emphasizing that the decision aid neither replaced specialist consultation, nor factored in personal medical history
Providers and Patients	• Leadership and key providers express support for the decision aid • General consensus that educational components of the decision aid are beneficial • Providers believe condition is preference-sensitive • Feedback of patient satisfaction scores and feedback of decision aid distribution data
BARRIERS	
Provider Decision Aid Issues	• Concern about the lack of clear evidence on best treatment option • Concern about the statistics and procedural risks cited in the decision aid • Lack of perceived decision aid need because of educational materials already available • Concerns about distribution of decision aid to inappropriate patients
Provider and Patient Issues	• Provider perceived difficulty in implementing decision aids and SDM with conditions considered by providers and/or patients as life-threatening • Negative patient reactions reported by providers • Lack of provider engagement
System or Condition Issues	• Difficulty identifying appropriate patients before visit • Difficulty scheduling time for follow-up conversations • Lack of formal service line leader/unfilled leadership position

First, the high level of leadership support at Group Health ensured that clinical staff received strong and consistent messages about the project's importance. Changes in organizational systems and procedures reinforced the expectation that the decision aids would be distributed to all appropriate patients. Also implied was the expectation that providers would engage patients in shared decision making conversations.

Second, developing a system for pre-visit decision aid ordering, when possible, appeared to improve overall distribution and create natural opportunities for patient follow-up. However, better methods for post-visit follow-up after decision aid delivery are needed to ensure that providers "close the loop" with patients before a final treatment decision is made, since providing the decision aid before consultation with a specialist is not always possible. Our evaluation findings suggest many missed opportunities when clinics rely heavily on providers ordering decision aids at the point of care. In these instances, patients either did not receive decision aids or did not have follow-up conversations with providers. Alternative distribution methods such as in-office viewing might ensure follow-up. However, these would require investments in infrastructure and staff time, and more patient time in-clinic.

Third, engaging providers in planning the distribution process was important. Even in supportive service lines that had high distribution rates, providers often had concerns about particular features of the decision aid content or distribution process. These were usually addressed by actively engaging providers in dialog regarding the video content and the distribution process. Service lines that engaged providers in preparation and planning for decision aid implementation had higher provider understanding and acceptance of the decision aids over time.

Fourth, decision aid distribution without provider engagement is not sufficient to facilitate shared decision making. For providers to change their conversations with patients, they must be included in conversations with leadership about the purpose of the decision aids and the types of changes expected. Most providers interviewed for this study felt that the decision aids were excellent educational tools, but failed to recognize their full potential as a tool for shared decision making. This helped convince leadership that training and peer support around shared decision making were needed so that providers understood what constitutes shared decision making and how to incorporate patient values in treatment decisions. As a result, a large-scale continuing medical education training for providers, solely on shared decision making, was planned for year three of the implementation process.

Finally, providers reported difficulty having shared decision making conversations with patients recently diagnosed with a life-threatening illness such as cancer or cardiac disease. This suggested that, for these life-threatening and time-sensitive conditions, providers require different types of engagement that increase their comfort in using the decision aids and engaging patients in shared decision making conversations.

Acknowledgements
The authors wish to thank the decision aid implementation team: Marc Mora, MD; Chris Cable, MD; Karen Merrikin, JD; Tiffany Nelson, Stan Wanezek, and Charity McCollum; the Research Advisory Panel: Michael Von Korff, ScD; Douglas Conrad, PhD; Carolyn Watts, PhD; Michael Barry, MD; Richard Wexler, MD; and Jeffery N. Katz, MD; and the other Group Health Research Institute employees who helped collect data and develop this manuscript: Sylvia Hoffmeyer; Carol Cahill MLS; Chris Tachibana, PhD; and Rebecca Hughes. We also thank Group Health clinical staff who made time to participate in the interviews and provide feedback on this manuscript. This research was funded by the Commonwealth Fund.

COMMENTARIES

Relational Coordination Theory: Jody Hoffer Gittell
Group Health's change process toward greater shared decision making was carried out in a highly favorable political context, with Washington State having redefined informed consent to include shared decision making. The change process also had a highly favorable organizational context, with top leaders within Group Health providing strong consistent messages about the importance of shared decision making. Even with this favorable context, the challenges involved in creating change were significant.

The primary innovation of decision aid distribution was integrating the decision aids into standard practice. A standard protocol was developed for identifying appropriate patients, then routing them into a new work process which included scheduling pre-visits to view decision aids. The use of process improvement helped expand distribution and engage staff and providers.

Distribution of decision aids, and even the effective use of decision aids by patients, did not however automatically translate into shared decision making between patients and providers. Providers continued to see decision aids as a form of patient education rather than as a basis for shared decision making. Providers remained uncomfortable with having these shared decision making conversations with patients, especially when those conversations involved life-threatening conditions, and thus were likely to be difficult or emotional.

Many case writers in this volume recommend additional training for providers. As noted above, Relational Coordination Theory suggests that this training could be designed explicitly as a *relational intervention*, an opportunity for care providers and patients to practice new ways of interaction to develop greater shared knowledge, shared goals, and mutual respect. In addition to this *relational intervention*, relational coordination theory suggests that successful change also requires *work process interventions*—changing the work process itself to incorporate shared decision making—as well as *structural interventions* reinforcing relational coordination and coproduction through changes in everyday structures for hiring, training, conflict resolution, performance measurement, rewards, protocols, meetings, and information systems.

Normalization Process Theory: Glyn Elwyn

The case from Group Health is a very good example of an organizational effort to normalize the use of patient decision support tools, at least in the 12 selected clinical pathways. In large part, normalization was achieved. The tools have become part of the way work is performed in Group Health. Elements of coherence, cognitive participation, collective action and reflexive monitoring are clearly visible within the work done by the organizational staff. There is also evidence that the work is extending to involve practitioners in primary care. Yet there are signs that normalization has only been achieved at an engineering level, with systems in place that are fairly successful at identifying eligible patients for these tools, and then achieving delivery. As the case writers admit, the impact of embedding the use of these tools in the face-to-face work of clinicians consulting with patients is unknown: a finding that is true, of course, for all the other cases in this book, as discussed in the introductory chapter.

Microsystems: Marjorie M. Godfrey

How the Group Health team designed, implemented, and evaluated their approach is an important contribution to understanding the long-term challenges of distributing decision aids into standard care practices. One of the project's unique characteristics is the geopolitical environment in which Group Health's practice lives. Washington State passed legislation supporting a demonstration program, believing fundamentally that the use of shared decision making is a higher form of informed consent and essentially providing the policy-level impetus to see change happen on the ground. In the case of Group Health, the decision aids, a type of decision support intervention, were distributed among specialty care groups as opposed to primary care units. What Group Health did well was to reach out to the specialty practices through two project management consultants (facilitators) to work with frontline providers and staff to solicit reaction and input into potential distribution pathways. This process contributed a sense of ownership and engagement to frontline providers and staff (microsystem).

Additionally there was strong senior leadership support (macrosystem) as demonstrated by the significant organizational resources committed to the project. Another strength to this model was the two-tiered process adopted for facilitating distribution of decision aids. Sixteen months into the project, senior leaders decided to adopt Lean management principles, resulting in major revisions to the decision aid workflows with a resulting optimization of the appointment schedulers' role. The decision aid ordering process was triggered by the scheduler's identification of key words and phrases patients might use while scheduling an appointment. There was an initial increase in the decision aid distribution to patients that decreased over six months. Modifying the electronic medical record streamlined the decision aid process and resulted in more decision aid distribution since providers role could now respond to electronic prompts. This demonstrates the role of technology in creating high-performing clinical microsystems. Process designs that include technology are most efficient, effective, timely, and patient-centered, while optimizing all roles in the practice.

Reflection on Commentaries

Group Health has dedicated significant resources toward maximizing the use of decision aids and increasing shared decision making for preference-sensitive procedures, in hopes that these efforts will reduce variation in practice and increase patient satisfaction. Group Health continues the work described in the case study by improving decision aid distribution mechanisms, training providers in shared decision making, and expanding decision aid offerings. As of January 2013, Group Health had distributed a total of 31,575 decision aids. The organization recognizes, however, that continued efforts are needed to sustain and build on successes to date. As the commentaries point out, Group Health's change processes touch on elements found in all three of the highlighted theories. The insights of these theories offer new opportunities for continued improvement to workflow and care processes aimed at promoting shared decision making at Group Health.

Clarissa Hsu, David T. Liss, Emily O. Westbrook, David Arterburn

Please direct permission requests for Chapter 5 to Clarissa Hsu at hsu.c@ghc.org

References

1. Kuehn BM. States explore shared decision making. *JAMA.* 2009;301(24):2539–2541.

2. Morgan MW, Deber RB, Llewellyn-Thomas HA, Gladstone P, Cusimano RJ, O'Rourke K, Tomlinson G, Detsky AS. Randomized, controlled trial of an interactive videodisc decision aid for patients with ischemic heart disease. *Journal of General Internal Medicine.* 2000;15(10):685–693.

3. O'Connor AM, Bennett CL, Stacey D, Barry MJ, Col NF, Eden KB, Entwistle VA, Fiset V, Holmes-Rovner M, Khangura S, Llewellyn-Thomas HA, Rovner D. Decision aids for people facing health treatment or screening decisions. Cochrane Database of Systematic Reviews. 2009;3(3):113.

4. O'Connor AM, Llewellyn-Thomas HA, Flood AB. Modifying unwarranted variations in health care: Shared decision making using patient decision aids. *Health Affairs.* 2004;Suppl Vari:VAR63–72.

5. Waljee JF, Rogers MAM, Alderman AK. Decision aids and breast cancer: Do they influence choice for surgery and knowledge of treatment options? *Journal of Clinical Oncology.* 2007;25(9):1067–1073.

6. Whelan T, Levine M, Willan A, Gafni A, Sanders K, Mirsky D, Chambers S, O'Brien MA, Reid S, Dubois S. Effect of a decision aid on knowledge and treatment decision making for breast cancer surgery: A randomized trial. *JAMA.* 2004;292(4):435–441.

7. O'Connor AM, Wennberg JE, Légaré F, Llewellyn-Thomas HA, Moulton BW, Sepucha KR, Sodano AG, King JS. Toward the "tipping point": Decision aids and informed patient choice. *Health Affairs.* 2007;26(3):716–725.

8. Hsu C, Liss DT, Westbrook EO, Arterburn D. Incorporating patient decision aids into standard clinical practice in an integrated delivery system. *Medical Decision Making.* 2013;33(1):85–97.

9. Graban M. *Lean Hospitals: Improving Quality, Patient Safety, and Employee Satisfaction.* Boca Raton: CRC Press, Taylor & Francis Group; Productivity Press; 2009.

10. Hadfield D, Holmes S. *The Lean Healthcare Pocket Guide: Tools for the Elimination of Waste in Hospitals, Clinics and Other Healthcare Facilities.* Chelsea: MCS Media Inc; 2006.

11. Hsu C, Coleman K, Ross TR, Johnson E, Fishman PA, Larson EB, Liss DT, Trescott C, Reid RJ. Spreading a patient-centered medical home redesign: A case study. *The Journal of Ambulatory Care Management.* 2012;35(2):99–108.

12. Womack JP, Byrne AP, Flume OJ, Kaplan GS, Toussaint J. Going lean in health care. In: Miller D, ed. *Innovation Series.* Cambridge: Institute for Healthcare Improvement; 2005:1–20.

CHAPTER 6

Step-by-step Using a Community Effort:

Implementing Patient Decision Support in Stillwater, Minnesota

Lawrence E. Morrissey Jr., Glyn Elwyn

Why we started: This case report describes efforts by the Stillwater Medical Group and the Minnesota Shared Decision Making Collaborative (MSDMC) to advance shared decision making in clinical practice. The Minnesota Shared Decision Making Collaborative was created to spread best practice in shared decision making, based on Minnesota's track record and infrastructure for community-wide quality improvement. A broad range of those in the health care system participated including: (1) patients, (2) health plans, (3) health care delivery systems, (4) purchasers of health care, (5) state government, (6) research institutions, (7) quality improvement organizations, and (8) the local professional association.

What we set out to do: We adopted an approach of "learning while doing," applying best available evidence within clinical practice. Leadership and staff support from the Stillwater Medical Group emphasized shared decision making by clarifying its value in providing patient-centered care, an organizational priority. At Stillwater an interactive, team-based process implemented shared decision making by developing pilots in selected preference-sensitive surgical conditions, like uterine fibroids and prostate cancer treatment, prior to expanding the program over three years.

What we achieved: Through surveys we measured changes in patient and provider experience and demonstrated successful implementation by measurement of decision support delivery rates and confirmation of patient sentiment that this was important to offer, thereby accomplishing change in a "real world" clinical setting. We see this as validation of continuing to expand shared decision making programs to other conditions while maintaining and improving the existing pilots.

What we achieved: We experienced many challenges to implementation including provider resistance to change, inconsistent delivery in certain conditions, and difficulty in getting patients to complete measurement tools in some conditions[1]. But the Minnesota Shared Decision Making Collaborative has definitely increased interest in shared decision making across the community and has inspired organizations to pilot shared decision making in clinical settings.

CASE REPORT

Why we started: The Stillwater Medical Group began work in shared decision making as part of their quality improvement programs, building on an aspiration to provide patient-centered care, an established core value of the organization. We recognized value in working with others to implement shared decision making and began working with Health Partners, another local, large, integrated health care delivery system. We also became a demonstration site in the Informed Medical Decisions Foundation (IMDF) research project on implementation. Our plan was to deliver patient decision support through use of specific decision aids, beginning with a pilot project for women with uterine fibroids and men with prostate cancer. We recognized an additional need for changes in organizational care delivery systems to support shared decision making throughout the care process. Finally, the Stillwater Medical Group played an active role in the Minnesota Shared Decision Making Collaborative and gained knowledge and support from other organizations, including Health Partners, the Mayo Clinic, and the Institute for Clinical Systems Improvement (ICSI).

What we set out to do: Members of the Minnesota Shared Decision Making Collaborative realized that we had mutual goals regarding shared decision making and so we held bimonthly meetings over the last three years, supported with staff and meeting space by Health Partners and ICSI. The Collaborative developed a state-wide initiative for implementing shared decision making, with the first step being development of a charter containing both objectives and a strategic plan to increase community awareness of shared decision making. The Charter's stated aims were:

- To identify best practices for implementing shared decision making and measuring decision quality;
- To implement and spread these best practices across Minnesota;
- To improve clinician-patient decision making, reducing or eliminating unwarranted variation in preference-sensitive care; and
- To create the structures and processes required to perform the work in sustainable fashion.

Specific measures tied to these goals included increasing the number of new pilots the collaborative could inspire and hosting a community shared decision making conference.

What we achieved: At Stillwater, work began by addressing women with uterine fibroids who had been referred to gynecologists. These patients were provided a decision aid (the Ottawa personal decision guide) and offered a phone number for nurse coaching after a visit with gynecologist[1]. Our data revealed inconsistent delivery of the tools. A total of 71 patients were eligible over a 12-month period; however, only 30 (42%) received the tools. Of these 30, 20 completed the survey and 17/20 viewed the tool and 14/20 used the accompanying booklet. However, when the tools were used by patients, evidence suggested that they were highly valued: according to the survey, all patients (n=18) felt that the tools should be offered routinely, 67% reported knowledge gain, and 73% felt that the tools helped them discuss treatment options with their clinician.

A simultaneous effort to build support for patient-centered care as a core organizational goal helped create shared knowledge about shared decision making. Visible leadership support for shared decision making and provider willingness to innovate allowed us to expand into our second pilot, for patients with prostate cancer. We delivered decision support to 100 of 122 eligible patients (82%), driven by an implementation process that delivered the materials before the provider visit using a decision support nurse, and highlighted by the involvement of a single urologist who felt strongly that shared decision making added value to his practice. Success in the prostate cancer pilot created further interest and enabled expansion toward a range of decision aids for other conditions (see Table 6.1), supported by a grant from the Informed Medical Decisions Foundation (IMDF).

Delivery rates of decision support have been inconsistent (see Table 6.1). We learned that it is vital to offer decision support at the right moment and that this varies significantly across clinical problems. In situations where diagnosis and the need for treatment decisions are clear-cut, e.g., in breast and prostate cancer, delivery is easier to accomplish. Our group also aimed to achieve what we call a "warm handoff," encouraging patients to access tools and reinforcing that we offer continued support. Where decision support delivery relies on the provider "remembering" or relies on the patient in some way, the rate of delivery is less consistent. This is consistent with known barriers noted in the literature[2].

TABLE 6.1
Delivery rates for selected conditions at Stillwater Medical Group

CONDITION	START	DECISION SUPPORT GIVEN	ELIGIBLE PATIENTS IN TIME PERIOD	DELIVERY RATE
Uterine Fibroids (restart after original pilot)	08/23/2010	27	116	23%
BPH	10/01/2009	118	791	15%
Prostate Cancer	06/01/2010	100	122	82%
Breast Cancer	11/02/2010	47	59	80%
Depression	06/15/2010	125	294	43%
Back Pain	06/27/2011	112	270	42%

Feedback from patients has been positive throughout and in surveys they were adamant that other patients would benefit from using these tools. They do note, however, that they need time to digest the new information contained in these tools. They also confirm that setting a clear expectation of expressing their personal preferences helps them engage in the process.

As for staff, 57 percent reported greater satisfaction when comparing their current approach to one used a year ago. After two years, employees gave high ratings (8.9 and 8.6 on a 0-10 scale) to the importance of identifying choices for patients and eliciting their preferences. Clinicians communicating the opportunity and expectation to participate has been identified as a potential facilitating factor for shared decision making[3]. Survey data doesn't capture the general positivity generated by the work; word-of-mouth testimonials praising implementation has helped break down barriers and encouraged new providers to try the process.

Another important if wider outcome has been the work of the Minnesota Shared Decision Making Collaborative, details of which can be found at:

msdmc.org. Working groups have provided: (1) definitions of shared decision making (Lexicon); (2) summaries of best practices for implementation; (3) strategies for measurement; (4) patient engagement; and (5) interaction with the mass media. The Collaborative has increased interest and supported other organizations to become engaged with research funding secured for focus groups to examine patient experiences (Stratis Health). There have been five pilot projects initiated and the Collaborative hosted a shared decision making conference. We feel the pilots demonstrated that there are important elements of shared decision making readily tailored to primary care[4].

What we learned: Implementation of shared decision making in clinical practice is difficult work. Delivery of shared decision making can be inconsistent, relying on the actions of individuals that result in less consistent delivery. The Stillwater Medical Group's process of gradual implementation enabled gradual expansion at a rate of change that the organization could tolerate. Yet despite the incremental ethos, or maybe because of it, commitment to shared decision making as an expression of patient-centered care is strongly supported by leadership and this organizational commitment has grown despite implementation challenges. The Group's efforts have resulted in a cultural shift towards viewing shared decision making as a core component of the care we deliver and we notice that these changes are happening at other organizations in our community as well, providing additional motivation. This model of a community-wide effort to use patient decision support and the gradual wider uptake that we have witnessed at Stillwater and in Minnesota could be replicated elsewhere.

Acknowledgements
Supported by a grant from the Informed Medical Decisions Foundation (IMDF).

COMMENTARIES

Relational Coordination Theory: Jody Hoffer Gittell
The Stillwater Medical Group leveraged Minnesota's history and infrastructure for community-wide quality improvement in its efforts to increase shared decision making. The SMG team achieved a range of positive outcomes—patients who received decision support tools valued them; staff were more satisfied relative to two years prior; and they were highly supportive regarding shared decision making's importance. But a large gap remained between patients who were eligible to receive decision support tools and those who actually received them, particularly in the gynecological setting relative to the prostate cancer setting.

How might we understand the successes and limitations of this effort from the standpoint of Relational Coordination Theory? In general, the Stillwater Medical Group's approach was highly consistent with the Relational Coordination Theory. They introduced shared decision making through the already-established quality improvement process used for many years, and thus aimed to "change care delivery systems to support shared decision making through the care process," creating supportive structures and processes as needed. The team used both work process interventions (redesigning the work process to focus on timing for delivery of decision aids) and structural interventions (changing staff roles to incorporate new responsibilities for supporting the shared decision making process) to make success less reliant on individual actions, and also less reliant on individual providers' ability to "remember."

Design of new staff roles differed, however, in the two pilot settings, particularly in the timing of staff engagement with the patient. In the gynecological setting, patients were offered time with a nurse coach after their visit to support use of the decision aid. In the prostate setting, patients received decision aids from a decision support nurse. This difference in timing may have contributed to 42% of patients receiving the tool in the first setting versus 82% of patients receiving the tool in the second setting. While both pilots included work process and structural interventions, their design in the prostate setting may have been more effective regarding timing.

Given the focus on delivering the decision tools, the case write-up does not reveal the extent to which the Stillwater Medical Group used relational interventions to change the dynamics of decision making between provider and patient. Data provided in the case do suggest strong overall cultural support for shared decision making, due in part to shared goals in the broader community for patient-centered care and for shared decision making. In effect, the relational intervention was at the community level, through the multi-stakeholder Minnesota Shared Decision Making Collaborative that included the voices of patients, health plans, health delivery systems, professional associations, and local government agencies. This is a broader relational intervention than the Relational Coordination Theory typically considers, providing an opportunity to expand to a new level.

Normalization Process Theory: Glyn Elwyn

As a demonstration site, Stillwater was part of the network supported by the Informed Medical Decisions Foundation (IMDF). When examined under the NPT lens, significant work was accomplished to achieve coherence. Many key leaders collaboratively developed a charter; coherence extended wider than the organization, as evidenced by the creation of the Minnesota Shared Decision Making Collaborative. Participants engaged in interactional work (cognitive participation) about who should do the work, or how do the decision tools get to the right patients' hands. They established agreements about how the work gets done (collective action) establishing systems to measure to what extent eligible patients were given decision support interventions (reflexive monitoring). The case also reveals that, despite evidence of achievements in each of the four NPT constructs, it was difficult to ensure that all eligible patients received the right tools, and in some clinical areas, coverage was low. Furthermore it is unclear whether the levels of implementation achieved have led to sustainable practice. Nevertheless, Stillwater provides a good example of how an organization might successfully develop a culture of using patient decision support and where NPT constructs are, at least to some extent, met.

Microsystems: Marjorie M. Godfrey

The Minnesota Shared Decision Making Collaborative, consisting of a broad spectrum of health care stakeholders at the state level, provides an exemplary example of how the "geopolitical" system can support health care delivery and improvement at the macro- and microsystem levels. Providing supportive tools and resources enabled the organization and front-line team to move forward with patient decision support. Occasional provider resistance to change and inconsistent delivery of patient decision support causes one to reflect, however, on the action of "implementation." Engaging the microsystem in the exploration of the various tools and resources, exploring their local work flow processes, cultural patterns, and inter-professional action might lead to adaptations of the process to build patient decision support into the care processes. Remembering that every microsystem is unique helps to remind anyone interested in health care improvement that inserting or installing change ideas often won't work. Engagement of everyone at the front line of care can lead to greater ownership and commitment to the redesign workflows and processes.

Disciplined improvement science requires review and identification of best practices and allocates time to identify good change ideas for new settings. Applied microsystem theory suggests that clinical care improvements require full support of the entire medical team including patients, clinicians, staff, and administration, represented in detail by the 5Ps: purpose, patients, professionals, processes, and patterns. In the case of Stillwater, there was wide clinical, environmental, and administrative support, yet no formal exploration of the 5Ps. From a microsystems perspective, Stillwater incorporated many core elements to making lasting process change possible without committing to a structured workflow plan. The group targeted patients, requested clinicians' feedback, and sought administrators' advice and support without formally committing to integrating each of these parts, thereby missing a key element to microsystem theory, where the patient and family are central.

The goal of offering decision support at the right moment should be explored with patients and families so as to learn from those with different diagnoses when, how, and by whom they should receive decision support. The fact the patients completed the surveys and strongly recommended decision support tools to other patients, while admitting the need for enough time to reflect on the decision aid information, further informs the work processes.

Reflection on Commentaries
The observations from the Relational Coordination Theory perspective, on how the clinic established optimal timing for offering decision support and moved towards processes that rely less on individuals, apply to all quality improvement efforts. Applying these concepts to organizations and to the community as a whole, are especially important. The symbiotic relationship between the work at the organizational level and the work at the community level enabled both projects to be "greater than the sum of their parts." When viewed through the lens of the Normalization Process Theory, shared decision making has become a more normal, more accepted concept as a standard of care. New organizations continue to explore its application and seek information about it. The microsystems analysis views the community developments as different groups finding their own way, working to define the most effective strategies, and accepting that there is not likely a "one size that fits all" solution, but rather a collection of viable options. This is a very helpful way to consider process implementation.

Lawrence E. Morrissey Jr.

Please direct permission requests for Chapter 6 to Lawrence E. Morrissey Jr. at lawrence.e.morrissey@lakeview.org

References

1. Solberg LI, Asche SE, Sepucha KR, Thygeson NM, Madden JE, Morrissey L, Kraemer KK, Anderson LH. Informed choice assistance for women making uterine fibroid treatment decisions: A practical clinical trial. *Medical Decision Making.* 2010;30(4):444–452.

2. Légaré F, Ratté S, Gravel K, Graham ID. Barriers and facilitators to implementing shared decision-making in clinical practice: Update of a systematic review of health professionals' perceptions. *Patient Education and Counseling.* 2008;73(3):526–535.

3. Edwards M, Davies M, Edwards AG. What are the external influences on information exchange and shared decision-making in healthcare consultations: A meta-synthesis of the literature. *Patient Education and Counseling.* 2009;75(1):37–52.

4. Murray E, Charles C, Gafni A. Shared decision-making in primary care: Tailoring the Charles et al. model to fit the context of general practice. *Patient Education and Counseling.* 2006;62(2):205–211.

CHAPTER 7

Orientation and Education:
Steps Needed to Get Clinicians to Order Patient Decision Support

Leigh Simmons, Karen Sepucha

Why we started: For several years primary care providers at Massachusetts General Hospital (MGH) have been able to order decision aids for their patients through the electronic medical record. However, use of this facility is highly variable. Some providers frequently order and others never order the use of decision aids. Barriers to consistent and frequent use included: (1) lack of familiarity with the program content; (2) concerns that the prescribing process was complicated and time-consuming, and (3) lack of familiarity with the prescribing program.

What we set out to do: We designed a course for clinicians to address these barriers. The one-hour continuing medical education (CME) course included viewing a decision aid, reviewing ordering procedures, and sharing data on peer presentation of decision aids at the individual and practice levels. We tracked orders and numbers of unique referring clinicians to evaluate the impact of the CME course on decision aid use. We also collected surveys from participants in the sessions.

What we achieved: We delivered the course to 15 primary care practices; 165 clinicians attended the sessions. Course feedback was positive, with a majority indicating they would change their practice. We noted a significant increase in orders after the course and a significant increase in the number of unique referring clinicians.

What we learned: When it comes to ordering decision support interventions for patients, clinician orientation to the referral model is an important component of programs designed to integrate shared decision making in routine care. Clinicians are more confident using decision support tools after they view and familiarize themselves with their content.

CASE REPORT

Why we started: The MGH Shared Decision Making Center tries to ensure that all patients who face a significant medical decision are well-informed about their options, involved in decision-making, and receive treatments that meet their goals and needs.

A core component of this work involves distributing patient decision aids for commonly encountered clinical conditions in primary care. The available decision aids are videos and booklets produced by the Informed Medical Decisions Foundation (IMDF), covering 35 common medical decisions. At the start of the project, an ordering system was embedded in the electronic medical record. Clinicians are able to order programs electronically, and video icons attached to items in the problem list remind them that a corresponding decision aid exists for those diagnoses. When the clinician places an electronic order, an email is sent to the MGH Blum Patient & Family Learning Center at the main hospital campus. The requested program is sent to the patient at home along with a questionnaire and a pre-paid return mailer.

The program was piloted in a primary care practice in 2005, and made available to 15 adult primary care practices affiliated with MGH by 2006. These practices serve more than 200,000 patients in Boston and its surrounding communities and are staffed by over 350 clinicians, including physicians trained in internal medicine and family medicine, internal medicine resident physicians, nurse practitioners, and registered nurses. The practices vary in specialty as some are hospital-based, others are community-based, and several focus on specific populations, e.g. geriatrics, women's health, and Hispanic populations.

Initially many primary care clinicians were excited about the potential use of decision support programs. Over time, however, most program orders came from a small set of physicians. In addition, only select decision support programs were ordered: namely, prostate cancer screening, colorectal cancer screening, and advance directives planning. These programs were featured in early promotion of the program within primary care practices, and reflected the decision aids clinicians most likely viewed, at least in part. In 2008, the project team surveyed all providers to identify facilitators and barriers to using decision aids in practice. The survey highlighted that clinicians who ordered decision aids were far more likely to have watched a program than those who had never ordered. Furthermore, those who had not ordered stated they would be much more likely to order programs if they were familiar with their content.

Similar barriers and facilitators have been found in other studies of decision aid use. For example, studies have found that physicians are more willing to use decision aids once given a chance to review them[1,2]. Gravel et al.[3] reviewed studies on decision aid implementation and found that the top three barriers reported by health care providers across studies included time constraints, lack of applicability of the decision aid due to patient characteristics, and lack of applicability due to clinical situation. Reviews also identified key facilitators to use of decision aids. The top three included provider motivation, perception of positive impact on clinical process, and perception of positive impact on patient outcomes.

What we set out to do: In recognition of these common barriers and facilitators and in response to specific clinician feedback, we developed a course to orientate and educate physicians about the ordering process and content of some of the tools. The literature on effectiveness of continuing medical education (CME) suggests educational sessions alone do not result in any behavior change: multiple methods are more effective[4]. Community-based strategies such as academic detailing (and, to a lesser extent, the influence of opinion leaders), practice-based methods such as reminders and patient-mediated strategies, and multiple interventions appeared to be most effective. Mixed results and weaker outcomes were demonstrated by audit and educational materials, while formal CME conferences without enabling or practice-reinforcing strategies had relatively little impact[4]. Thus, we designed our CME intervention to be more than a formal presentation on shared decision making.

The course aimed at giving primary care clinicians an opportunity to view a decision aid in its entirety. We gave them opportunity to discuss some of the challenges with decision aids and shared decision making with program leadership as well as with their practice colleagues. They also learned more about the electronic ordering program. Each practice director could choose one of three decision aids for viewing during the session: for colon cancer, knee osteoarthritis, or a general audience program on advance directives planning. Most practices chose to watch the program on knee osteoarthritis, a relatively new decision aid that few clinicians had viewed or ordered for patients. The course also provided an opportunity to share provider and practice level data on usage, with comparative data offered to increase motivation for use of the programs (see Figure 7.1 for an example of the performance data). CME credits were available for physician attendees.

At the end of the course we collected feedback in the form of a short written survey. We also examined its impact by comparing prescription rates in

FIGURE 7.1
Performance feedback comparing prescription rates across practices

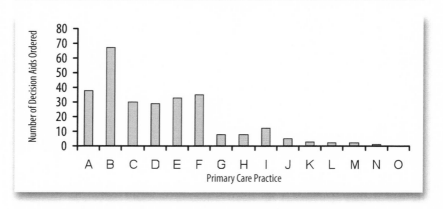

the months before and after the course in a statistical process control (SPC) chart. The SPC chart plots the mean, actual values, and upper and lower limits that indicate the bounds above and below which the process is considered "out-of-control." A grouping of seven data points above or below the mean indicates a statistically significant change in the process, so we looked for improvements due to the CME course[5]. We also examined the number of unique prescribers in the practice over the four weeks prior to the course compared to the four weeks afterwards.

What we achieved: The one-hour course was delivered to the 15 primary care practices from October 2010 through May 2011. The courses were attended by 165 clinicians, including physicians (the majority of attendees), nurses, nurse practitioners, and medical assistants. Most courses took place during regularly scheduled team meetings. (See Figure 7.2 for the impact of the courses on overall prescription rates.) The mean monthly prescription rate was about 100 before the interventions started and increased to about 250 after all CME interventions were conducted. The number of unique providers who prescribed at least one program increased as well, from 40 to 65 in four weeks pre-intervention compared to four weeks post-intervention, p<0.001.

Clinicians who attended the sessions reported a positive experience. Most indicated that the session was valuable, though some expressed lingering reservations about implementation of patient decision aids in their practice. The majority of respondents, 103/120 (86%), rated the session as useful or very useful. Eighty-eight percent of respondents stated they would make changes in their practice based on the session. Some comments from the written surveys illustrated the session's benefits:

> *"After watching the video, I know what my patients will be seeing"*
> *"I didn't realize the prescriptions were so easy to do"*
> *"[This session] made me aware of the breadth of available programs"*
> *"This [comparative practice data] spurs me on to prescribe!"*

Comments also reflected concerns about ability to put methods into practice:

> *"Will all patients have access to video players?"*
> *"I doubt my patients will be willing to view a full video"*
> *"[Low] education, literacy, cultural disparities for our multicultural, immigrant community will be a barrier to use of the programs"*

Despite these concerns, providers increased use of the programs and many new providers started using them. (Figure 7.2 demonstrates that the initial increase in prescriptions was sustained across practices over subsequent months.)

FIGURE 7.2
Training intervention significantly increased number of decision aids ordered

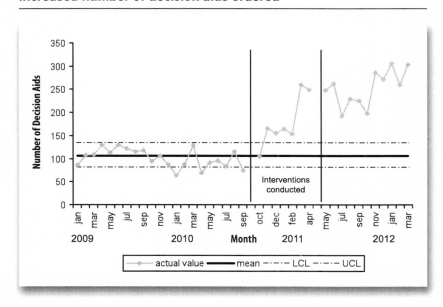

What we learned: The short instrumental and educational intervention of the CME course had a significant impact on clinicians' use of decision aids and led to increases in the number of decision aids ordered as well as to the number of prescribers. These data are reinforced by the positive physician evaluations and high numbers who said they would change their practice based on the session. It appears that a single short session could address multiple barriers to prescribing decision aids, including concerns about program content, worry that prescribing videos is complicated and time-consuming, and unfamiliarity with the program.

There are some limitations with the approach and measurement. First, we did not randomly assign practices to receive the training or not, and it is unclear how the different components of the multifaceted program contributed to the results. A recent Cochrane systematic review of approaches to increase clinician adoption of shared decision making did not find much evidence of sufficient quality to review[6]. It will be important to balance the need for rigorous evaluation of different approaches to implementation with the need for rapid turnaround and flexibility often required for successful quality improvement projects[5].

MGH's program in shared decision making is poised to deepen and expand. Providing decision aids is key to ensuring that patients are prepared for decision making discussions. Equally, or perhaps more important for widespread implementation of shared decision making will be training clinicians in the use of decision aids and in skills for shared decision making. Using video decision aids as a teaching tool for clinicians (e.g., to illustrate language that patients use to describe topics, and how to convey risk and uncertainty in terms that patients understand) must be explored further. In fact, one practice approached us after their session to continue the viewings. They have now watched at least four other programs during team meetings. We are tracking whether the viewings increase orders for the viewed programs and more broadly across the library of available programs.

We have continued to provide quarterly feedback reports on prescription rates to all our practices. In addition, we are considering additional clinician training to cover issues such as how to introduce decision aids at the optimal time, how to conduct follow-up conversations after using decision aids, and how to have shared decision making conversations in situations where no patient decision aids are available. Finally, we have designed and delivered a series of education sessions for internal medicine residents that covers shared decision making skills. These skills will be a core CME component and this program has served as a solid launching point for ongoing educational initiatives in shared decision making in primary care.

COMMENTARIES

Relational Coordination Theory: Jody Hoffer Gittell

This change process at MGH focused on increased use of decision support tools in the primary care setting. The main intervention was training primary care teams about the content of decision support tools and the process for prescribing them, to overcome lack of clarity and lack of familiarity, and ultimately to increase the rate of prescribing tools to eligible patients. After the training, there was a substantial increase in the number of primary practices that prescribed decision support tools, and a substantial increase in the number of decision support tools prescribed per practice. The training was clearly successful in achieving its goals.

The training was not really designed to increase either relational coordination or relational coproduction. There may have been an increase in relational coordination, specifically an increase in shared knowledge among care providers due to greater understanding of how the tools are intended to work. But it is unclear whether the training had any impact on relational coproduction, i.e., staff engagement with patients in the process of shared decision making. To achieve changes in shared decision making, the authors acknowledge, may require a different kind of training.

Relational Coordination Theory suggests that follow-up training could be designed explicitly as a *relational intervention*—an opportunity for care providers and patients to practice interacting in new ways to develop greater shared knowledge, shared goals, and mutual respect. In addition to this relational intervention, RCT suggests that successful change also requires *work process interventions* (changing the work process itself to incorporate shared decision making); as well as *structural interventions* (reinforcing relational coordination and coproduction through changes in everyday structures for hiring, training, conflict resolution, performance measurement, rewards, protocols, meetings, and information systems).

Normalization Process Theory: Glyn Elwyn

Familiarity with the content led to increased provider willingness to prescribe decision support to patients. This was the clear finding at MGH, which also provided instruction about how providers should use the electronic system to send the tools to patients. Viewed from the perspective of NPT, this seems to be a classic example of ensuring coherence and cognitive participation. Making sense of the intervention (knowing more about it), and knowing more about what is actually involved in ordering decision aids, led to documented behavior change. For complete normalization, practitioners would be expected to assume the task of ensuring that the work gets done, that the relevant patients are given the correct tools, and that performance against their own standards is monitored. The case does not suggest that practitioners achieved this level but clearly indicates that availability of patient decision support tools is insufficient motivation for change.

Microsystems: Marjorie M. Godfrey

The MGH team has a long and well-established history of using decision support in primary care aided by the electronic medical record for ordering decision aids. The use of the electronic medical record highlights the intentional support of the macrosystem and the mesosystem for use of technology to ensure microsystem success. A supporting microsystem, the MGH Patient and Family Learning Center, participates in the workflow by distributing the decision aid programs as requested through the electronic medical record. MGH reports that despite integration into the electronic medical record, use of decision tools varied. They identified providers as unfamiliar with program content and lacking understanding of the prescribing process and prescribing program. In light of these issues the MGH team tested a new model to facilitate provider interest and ultimately use of decision aids.

Their approach focused primarily on education of nurses, physicians, nurse practitioners, and medical assistants. Activated clinical microsystems include all members of the microsystem, and based on this report, it doesn't appear that support and clerical staff or patients were involved in the educational processes. The course feedback and data charted by using control charts shows empirically the value of these courses to increase knowledge and overall use and referral of decision aids. Also noteworthy: besides education about the decision aids and use of the electronic medical record to prescribe decision aids, there didn't appear to be any effort to convene the interprofessional practice to discuss and explore current work processes and timing of decision aid prescription.

Additionally, there did not appear to be discussion of processes for follow-up with patients who received the decision aids or about what the process looked like in the absence of a decision aid. High-performing microsystems demonstrate a deep understanding of core care processes. How decision aids are triggered, and identifying who does what and when, ensures reliability and is important work for every member in the practice including patients. The inclusion of medical residents is noteworthy, since educating them is an important process within patient-centered care. The MGH team recognized the importance of long-term sustainability by providing quarterly performance feedback to the microsystem to encourage the practice of decision aid use and to maintain awareness of current performance for the sake of making real-time adjustments.

Reflection on Commentaries

We appreciate the commentaries which have provided several concrete ideas for ways we might improve upon our processes. Since this work was presented, we have started to focus more on relational interventions, and have some experience with a new curriculum for residents that incorporates role plays where they practice shared decision making conversations. Ideally, we would like to bring patients to these sessions to teach and learn along with our clinicians. We also agree that it is important for us to do more to engage office staff, who are often the people ordering the decision aids, into training sessions and into workflow design discussions.

Leigh Simmons, Karen Sepucha

Please direct permission requests for Chapter 7 to Leigh Simmons at lhsimmons@partners.org

References

1. Feibelmann S, Yang TS, Uzogara EE, Sepucha KR. What does it take to have sustained use of decision aids? A programme evaluation for the Breast Cancer Initiative. *Health Expectations.* 2011;14 Suppl 1:85–95.

2. Graham ID, Logan J, O'Connor AM, Weeks KE, Aaron S, Cranney A, Dales R, Elmslie T, Hebert P, Jolly E, Laupacis A, Mitchell S, Tugwell P. A qualitative study of physicians' perceptions of three decision aids. *Patient Education and Counseling.* 2003;50(3):279–283.

3. Gravel K, Légaré F, Graham ID. Barriers and facilitators to implementing shared decision-making in clinical practice: A systematic review of health professionals' perceptions. *Implementation Science.* 2006;1:16.

4. Davis DA. Does CME work? An analysis of the effect of educational activities on physician performance or health care outcomes. *International Journal of Psychiatry in Medicine.* 1998;28(1):21–39.

5. Langley GJ, Moen RD, Nolan KM, Nolan TW, Norman CL, Provost LP. *The Improvement Guide: A Practical Approach to Enhancing Organizational Performance.* 2nd ed. San Francisco: Jossey-Bass Publishers; 2009.

6. Légaré F, Ratté S, Stacey D, Kryworuchko J, Gravel K, Graham ID, Turcotte S. Interventions for improving the adoption of shared decision making by healthcare professionals. *Cochrane Database of Systematic Reviews.* 2010;5(5).

CHAPTER 8

Implementing Tailored Decision Support Using a Patient Health Survey

Carmen L. Lewis, Shaun McDonald, Alison Brenner, Matthew Waters, Cristin Colford, Kim Young-Wright, Robert Malone

Why we started: Patient decision support has been shown to increase patient satisfaction and patient participation in office visits. However, significant barriers exist to implementation in primary care practices.

What we set out to do: We designed the Patient Decision Quality Initiative as a comprehensive patient decision support delivery system. For this phase we wished to identify primary care patients who could benefit from decision support interventions and then deliver them at the point of care. At the same time, we leveraged changes made in the clinical information system and workflow to assess patients' use of decision support.

What we achieved: We developed an algorithm to offer patients decision support interventions based on their known conditions and reported symptoms, as determined through questions on a patient health survey. 177 patients requested that they receive a decision support intervention through the patient health survey. During this same period, 364 patients returned for a follow-up appointment, and nurses completed the follow-up session for 266 (80%). Seventy-nine percent of the 200 patients who remembered receiving decision support reported having used it.

What we learned: We could modify our clinical information system and nurse workflow to successfully deliver tailored patient decision support for multiple primary care topics. We also were able to measure use of decision support using the same processes. Applying quality improvement methodology to decision support implementation shows promising results, although significant challenges remain.

CASE REPORT

Why we started: Patient-centered care is a measure of high-quality medical care[1]. One way to achieve patient-centered care is to provide patient decision support, especially since the use of patient decision support interventions has been shown to increase patient satisfaction and patient participation in office visits[2]. Significant barriers exist to implementation in primary care practices, however, including lack of time and reimbursement for these interventions, existing practice norms, and patient expectations. In previous work at the University of North Carolina we tested two delivery models to overcome these barriers. In the first we delivered decision support interventions before a scheduled office visit[3]. We mailed decision support for colorectal cancer screening in both DVD and video format to unscreened patients in our practice who were scheduled for an upcoming visit. Although this model had high reach (i.e., all unscreened patients in our practice) the proportion who reported having viewed some or all of the video or DVD was only 20 percent.

We then developed and tested an in-clinic delivery model to try to increase patient uptake. In this model, a staff member identified potentially eligible patients for weight loss surgery and prostate specific antigen (PSA) testing decision support[4]. The staff member used a portable DVD player to deliver the appropriate patient support intervention. We hypothesized that encouragement by a practice staff member would increase uptake compared to mailing decision support before scheduled visits. Although uptake improved to 58% we were able to reach only 59% of patients we had identified as potentially eligible. From this work we concluded that to maximize both reach and uptake required a more comprehensive approach.

What we set out to do: We designed the Patient Decision Quality Initiative as a comprehensive patient decision support delivery system, with the overarching goal to deliver "appropriate decision support to eligible patients all of the time." Our objectives were to: (1) create a sustainable system for delivery of decision support in our practice; (2) automate delivery as much as possible; (3) use continuous quality improvement methods to integrate decision support services into workflow; and (4) create cultural change for patient-centered care among staff, patients, and physicians.

As is customary for quality improvement projects, we performed numerous Plan-Do-Study-Act (PDSA) cycles[5]. This allowed us to develop and test a variety of processes, modify these processes to achieve the most effective methods, and disseminate those that reached targets. For this particular phase, we wanted to determine if we could identify primary care patients at the point of care who could benefit from decision support for a variety of conditions. To accomplish this we modified both our clinical information system and nurse workflow so that nurses delivered a patient health survey when they checked in patients.

The questions were tailored to the patient's known conditions as determined by billing or clinical data. Patients were also asked about specific symptoms. The patient health survey also assessed patient interest in obtaining decision support for the particular symptom or condition. If patients requested information, a relevant decision support tool was delivered during the office visit or they could request it to be mailed to their home. The delivery information was captured in our clinical information system, so that follow-up could be performed when they returned for their next visit.

At the same time, we leveraged these changes in the clinical information system and workflow to assess use of decision support in patients already receiving decision support. Nurses asked whether the patient had used the decision support and recorded their satisfaction. This uptake information was also captured in our clinical information system.

What we achieved: We developed an algorithm to offer patients decision support interventions based on their known conditions or common symptoms for which decision support was available (see Figure 8.1).

FIGURE 8.1
Clinical information system delivery flow-chart

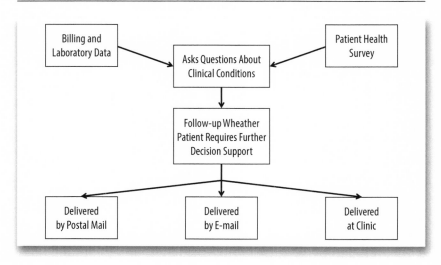

This project was funded by the Informed Medical Decisions Foundation (IMDF)[6], and so for this project we were implementing primary care topics that they had developed. The decision support interventions were DVDs that lasted 30 to 40 minutes and were accompanied by a booklet. The algorithm first presented questions that were symptom-based, as we wanted to prioritize symptomatic patients for decision support, thinking it would lead to increased uptake. If patients did not report current symptoms, they were then asked a question to confirm whether they had the chronic condition identified by the clinical information system, and were offered decision support for this condition. If the patient endorsed a symptom or condition, several follow-up questions asked if the condition was bothersome enough to warrant receiving further decision support. If they responded positively they were asked whether they wanted the DVD to be delivered while they were in the practice or mailed to them instead.

Decision Support Delivery: During this phase (August-December 2010), 177 patients requested that they receive a decision support intervention through the patient health survey. (The topics and distribution are shown in Table 8.1.) All six decision support interventions were requested by patients with a fairly even distribution, with hip and knee osteoarthritis being the most common.

Measurement of Decision Support Use: During this same period, 364 patients returned for a follow-up appointment after having previously received a

TABLE 8.1
Number and types of decision aids distributed

DECISION AID	NUMBER
Benign Prostate Hyperplasia	8
Colon Cancer Screening	16
Chronic Pain	26
Diabetes	26
Hip Osteoarthritis	46
Knee Osteoarthritis	38
Menopause	9
Weight Loss Surgery	8

decision support intervention. Nurses asked follow-up questions for 91% of these patients and completed the questions for 80% of those who began the patient health survey. Among these patients 75% remembered receiving decision support and 74% of those patients watched or read some or all of the DVD or booklet. Of the 200 patients who recalled receiving the decision aid, 178 reported satisfaction with the program (89%) and 179 considered this type of education useful (90%).

What we learned: We were able to modify our clinical information system and nurse workflow to successfully deliver tailored patient decision support for multiple primary care topics. We also were able to measure use of decision support using the same processes. All this was well received by patients. Most patients reported using the decision support interventions and most reported satisfaction with them.

Applying quality improvement methodology to decision support implementation shows promising results, but significant challenges remain. Funding for this project was time-limited and although changes in the clinical information system are enduring, monitoring and revisions are continually necessary. Appropriate resources will still be needed to oversee and revise work processes because of the dynamic nature and competing priorities of primary care. New US requirements emphasizing patient-centered care and electronic medical records[7] could potentially contribute to decision support implementation using both changes in work process and clinical information system, as we demonstrated with this project. Health care systems that invest in decision support implementation could benefit with enhanced patient allegiance, improve clinical decisions, and quality of patient care.

Acknowledgements
Supported by a grant from the Informed Medical Decisions Foundation (IMDF).

COMMENTARIES

Relational Coordination Theory: Jody Hoffer Gittell
After observing relatively low uptake of decision support tools in primary care (20% when mailed ahead, 58% when watching with staff), this University of North Carolina (UNC) program employed three interventions to increase the use of the tools. The first intervention sought to change information systems to deliver decision support. The second intervention introduced new measures of decision support and embedded them into the information system. The third intervention was changing the work process itself to

incorporate delivery of decision support using QI methods, specifically the Plan-Do-Study-Act (PDSA) cycle. With these interventions, use of decision tools increased to 74%.

Although these interventions included both *structural interventions* and *work process interventions*, as is consistent with Relational Coordination Theory, they did not include explicit attention to relational interventions. Moreover, all interventions focused on the increased use of decision tools rather than an increase in shared decision making—a common misunderstanding. Although we can hope that decision support tools lead to shared decision making, it is arguably true that shared decision making requires more fundamental changes in current patterns of interaction between care providers and patients (and among care providers themselves), and that these changes involve development of shared goals, shared knowledge, and mutual respect, supported by timely, accurate, problem solving communication. Changing these patterns to support shared decision making would likely require *relational interventions* that create a safe space for trying out new patterns of interaction, supported by tools such as relational mapping and interdependence conversations.

Normalization Process Theory: Glyn Elwyn

Analyzing the case by using the NPT presents a challenge. The barriers to the use of patient decision support interventions are familiar and are described by many similar cases here. There is a lack of time in clinics for such interventions, not to mention low motivation and no clear rewards to reinforce a change in practice. NPT would provide a set of analytical lenses to examine these issues, i.e., its four constructs of coherence, cognitive participation, collective action, and reflexive monitoring. However, the intervention here was not targeted at professional staff. It was targeted at *patients*; i.e., they were sent a survey that in the course of asking questions about health care also alerted them to the existence of tools that might interest them. This generated more interest in receiving the tools. In other words, it generated a patient "pull" instead of relying on the usual professional "push." It seems that some success was achieved, although details on numerators and denominators are sketchy. NPT does not provide a good framework for considering the contributions of other actors such as patients in creating demand for an intervention. Questions remain as to whether using a patient survey led to increased coherence, cognitive participation, collective action, and reflexive monitoring among the professional team. There are hints that it did not and that sustainable change was not achieved by this method.

Microsystems: Marjorie M. Godfrey

The UNC team aimed to improve the delivery of quality primary care using tailored patient decision support. This project focused its attention primarily on the clinical microsystem members such as nurses, staff, and patients, and also on workflows, technology, and patient-centered preferences. The team designed its actions based on two earlier improvement efforts (Plan-Do-Study-Act cycles) to inform their next PDSA cycle. This disciplined improvement approach is characteristic of a high-performing microsystem; standardized improvement including data to inform the improvement is an essential quality of microsystem professionals. The UNC team also studied patients in their practice to determine what mattered most to the various subpopulations. They designed a survey to assess patient interest and need before creating a process to offer decision support. We learned that exploring patient populations more deeply as a means of customizing a new care delivery model resulted in better matching of microsystem resources to patient needs.

A principle of high-performing microsystems is that they decrease variation in care processes while at the same time customizing patient care. This is clearly evident in the team's efforts to customize the survey questions based on patient condition. Collaboration at the mesosystem level through the use of billing and clinical data sources helped identify patient subpopulations and supported the microsystem improvement efforts with information technology. The team's report suggests that some authority and leadership support beyond the microsystem was garnered for access to information about patient records and appointments. The team acknowledged potential limits to the sustainability of this project. But they should celebrate their knowledge and discipline in quality improvement! Paul Batalden, pioneer of the microsystems approach, always emphasized that health care professionals have a responsibility to provide and improve care, a professional value very well demonstrated here. Continued monitoring of outcomes for patient decision support in the primary care practice, using both quantitative and qualitative data to report successes and challenges to meso- and macrosystem leaders, increases transparency of improvement efforts. Leaders throughout the organization will appreciate the "managing up" accomplished by the communicating of needed resources from the primary care practice along with supporting patient-centered care data.

Reflection on Commentaries

The goal for this project was to demonstrate whether we could systematically implement decision support tailored to individual patients' medical conditions and preferences. To provide uptake information to providers at the point of care and for sustainability, we measured whether patients used the tool as a part of the delivery system, which as we expected, resulted in some loss of follow up information because nurses were interrupted or had higher priorities during the visit. Of course, to promote, achieve, and measure whether shared decision making occurs after decision support will require interventions focused on providers and interactions between patients and providers. Now that we have a reliable delivery system in place, we hope to evaluate elements of shared decision after visits and provide training in shared decision making to our housestaff.

Carmen L. Lewis, Shaun McDonald, Alison Brenner, Matthew Waters, Cristin Colford, Kim Young-Wright, Robert Malone

---

Please direct reference requests for Chapter 8 to Carmen L. Lewis at carmen_lewis@med.unc.edu

References

1. Committee on Quality Health Care in America, Institute of Medicine. *Crossing the Quality Chasm: A New Health System for the 21st Century.* Washington, DC: National Academy Press; 2001.

2. Stacey D, Bennett CL, Barry MJ, Col NF, Eden KB, Holmes-Rovner M, Llewellyn-Thomas HA, Lyddiatt A, Légaré F, Thomson RG. Decision aids for people facing health treatment or screening decisions. *Cochrane Database of Systematic Reviews.* 2011;10(10).

3. Lewis CL, Brenner AT, Griffith JM, Pignone MP. The uptake and effect of a mailed multi-modal colon cancer screening intervention: A pilot controlled trial. *Implementation Science.* 2008;3:32.

4. Miller KM, Brenner AT, Griffith JM, Pignone MP, Lewis CL. Promoting decision aid use in primary care using a staff member for delivery. *Patient Education and Counseling.* 2012;86(2):189–194.

5. Langley GJ. *The Improvement Guide: A Practical Approach to Enhancing Organizational Performance.* 1st ed. San Francisco: Jossey-Bass Publishers; 1996.

6. Informed Medical Decisions Foundation. Informed Medical Decisions Foundation [website]. http://informedmedicaldecisions.org/. 2012.

7. Blumenthal D, Tavenner M. The "meaningful use" regulation for electronic health records. *The New England Journal of Medicine.* 2010;363(6):501–504.

CHAPTER 9

Integration Challenges:
Loss of an Existing Patient Decision Support System When a New Electronic Health Record Arrives

*Kate Clay, Stuart W. Grande, Carol DuBois,
Ivan Tomek, Melanie P. Mastanduno*

Why we started: The Dartmouth-Hitchcock Medical Center is an academic medical center that has been home to the Center for Shared Decision Making since 1999. The Center for Shared Decision Making was created to support patients in making good health care decisions, including redesigning clinical pathways to incorporate the use of decision aids as standard practice.

What we set out to do: In 2005, *Phase One* was initiated with the goal of integrating decision support into the care pathway for all eligible patients with a diagnosis of knee or hip osteoarthritis. In 2010, *Phase Two*, the transition to a commercial electronic health record required transforming this particular specialty care pathway through multiple workflow assessments and redesigns. The redesign effort included the feedback and active participation of surgeons, physician assistants, nurse practitioners, other clinical and administrative staff, and most notably a patient and family advisor.

What we achieved: *Phase One* measured and collected data about decision quality, decision process, and decision aid acceptability. In *Phase Two*, an ongoing three-year electronic health record transition and system change, data continues to be collected for the following variables: provider adoption, patient satisfaction with the electronic mode of administration, and reliability of patient information from survey responses through the physician visit.

What we learned: System changes required by the new electronic health record invalidated existing data collection systems, making replication of that data collection process by the new system more difficult. While challenges transferring processes from one medical record system to another may be unavoidable, minimizing future barriers to electronic health record implementation must include a multidisciplinary team, debriefing, process revisions, and a commitment to participating in further cycles of debriefings, revisions, and evaluations.

CASE REPORT

Why we started: The Dartmouth-Hitchcock Medical Center (DHMC), an academic medical center in rural central New Hampshire, is home to the Center for Shared Decision Making, both with high goals for patient care, research, and education. Through a research agreement with the Informed Medical Decisions Foundation (IMDF) and Health Dialog, Inc., the Center for Shared Decision Making focused on integrating the use of video decision aids into clinical care and studying their impact on patient decision making. As a strategic partner of the Dartmouth Institute for Health Policy and Clinical Practice (TDI), DHMC is similarly motivated by a commitment to health care improvement, improved patient outcomes, and process efficiency, prompting orthopaedic leadership at DHMC to champion use of decision support in the clinical setting[1,2]. Five of the early decision aids available to patients concerned orthopaedic specialty care: chronic low back pain, lumbar herniated disc, lumbar spinal stenosis, knee osteoarthritis, and hip osteoarthritis.

The department of orthopaedics at DHMC has 22 attending physicians and a post-graduate residency in orthopaedic surgery currently training 24 residents. Half the residents complete a master's level program at TDI which includes extensive training in biostatistics, epidemiology, health care quality improvement, systematic review, public health theory, and geographic health care practice variations[3]. Due to a heightened awareness of the importance of informed patient choice, the Center for Shared Decision Making acknowledged that working with the Spine Center and its Adult Total Joint Replacement clinic was a natural fit. The department aimed to move beyond clinical process barriers and integrate decision support into standard care for patients with knee osteoarthritis who were eligible for total joint replacement.

What we set out to do: In 1999, under the leadership of Dr. James Weinstein and Dr. John Wennberg, the Center for Shared Decision Making began operating as a clinical laboratory, with one of its principal goals the identification of clinical units with the greatest potential for integrating video decision aids into usual care. Within DHMC's Spine Center, the orthopaedics department gained experience with providing decision aids and gathering patient self-reported data for back conditions. This served as a valuable prototype for introducing decision support into workflow processes in a second orthopaedic sub-specialty.

In 2005, the Center collaborated with two total knee joint replacement surgeons to begin what turned out to be a two-phase project with the goal of integrating decision support into usual care for patients presenting with knee osteoarthritis. The initial phase of integrating decision support (with decision aids and decision quality questionnaires) was accomplished in the setting of a home-grown electronic health record which interfaced with a privately contracted patient-reported health questionnaire through a patient Web portal. The second phase became necessary when the home-grown medical record was replaced with a commercial electronic health record product.

Phase One: The aims were, first, to demonstrate the feasibility of integrating decision support and patient-reported decision making and functional status for knee osteoarthritis into the total joint replacement clinic workflow; and, second, to sustain decision support in usual care at the total joint replacement clinic after the research phase was concluded.

The four steps of Phase One were: (1) diagramming existing workflow for patients with end stage osteoarthritis of the knee (see Figure 9.1); (2) incorporating the patient triage algorithm used by scheduling secretaries to match the right patient with the right clinician at the right time; (3) mapping the high-level workflow to reflect the total patient care pathway for patients choosing surgery; and (4) identifying the step in the process when a surgeon determines a patient's eligibility for surgery and incorporating clinician prescription of the decision aid *Treatment Choices for Knee Osteoarthritis*. For patients, viewing the decision aid was paired with a pre- and post-viewing questionnaire that identified degree of knowledge of important facts, personal values, and current treatment choice. An electronic intake questionnaire completed by patients before the clinic visit fed forward some of this information to the clinician (see Figure 9.2).

Phase Two: The second phase began in 2010 and involved streamlining deployment of the decision aid utilizing the new electronic health record technology from the vendor EPIC. Phase Two's aim was to enhance the existing decision support process by incorporating the knee osteoarthritis decision aid and patient-reported decision making data into the standard processes embedded within the new electronic health record framework designed for the total joint replacement clinic. Enhancement of the existing process came via an electronic trigger that prescribed the decision aid so patients viewed it prior to the initial visit with the total joint replacement surgeon. The electronic trigger was paired with an electronically-collected self-reported decision quality information aid at multiple time points that was available for review by

FIGURE 9.1
Proposed decision making process for total knee replacement

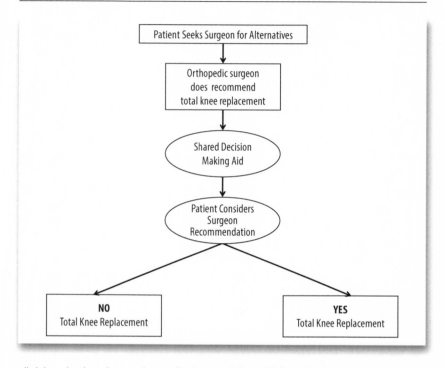

clinicians in the electronic medical record. A multidisciplinary team of surgeons, physician's assistants, nurse practitioners, clinical and administrative staff, and a patient and family advisor engaged in six months of planning, meeting weekly to map current workflow, draft new workflow, participate in facilitated simulation of the new workflow, streamline the e-survey process and questions, assist in testing, and prepare for deployment of eD-H (the new certified electronic health record system at DHMC).

For the clinic and the development team, successful implementation needed to include: (1) a seamless transition from previous electronic health record technology; (2) incorporation of the multidisciplinary team's recommendations and proposed "ideal" framework; and (3) new functionality that supported the clinical team's intentions to systematically and automatically deliver the decision aid to the right patients at the right time.

What we achieved: *Phase One:* After completing Phase One, patients seen in the total joint replacement clinic received clinician prescriptions for the

FIGURE 9.2
Decision support integration

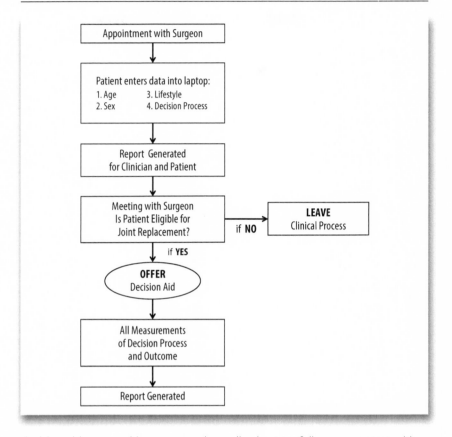

decision aid, accessed in person or by mail prior to a follow-up contact with the clinic, thus closing the decision making loop. We were able to measure and analyze patient characteristics on decision quality, decision process, and decision aid acceptability through use of a paper questionnaire which accompanied the decision aid. Some of this data was provided to physicians for use at the point of care.

Data collected by the Center for Shared Decision Making from the first phase included nearly 3,000 hip and knee osteoarthritis decision aid loans to patients, of whom 38% (n=976) returned the decision quality questionnaires, demonstrating the decision aid's impact regarding knowledge, choice certainty, treatment preference switching, and acceptability.

Phase Two: Following a three-year electronic health record transition from a home-grown system to a commercial product, only some Phase One accomplishments were sustained. Patient data from eD-H continues to be collected for the following variables: provider adoption of decision aids, patient satisfaction with the electronic decision aid interface, and reliability of patient information from survey responses through the physician visit. Right now, virtually all designated patients receive the electronic questionnaires before their visit. Outstanding questions following the transition include: How reliable is the eD-H tool for finding the right patient at the right time? What is the best way to validate the patient's history and its consistency with the intake e-questionnaire? How will the patient's decision support feedback and data be accessible and easily visible to the physician at the point of service?

What we learned: As outlined previously, the transition to eD-H was not as seamless as originally hoped. The system is not yet able to initiate a decision support referral (an electronic trigger to mail the decision aid/decision quality questionnaire to patients scheduled for knee pain evaluation) to inform patients of care options prior to their initial surgeon's visit. Ideally, the data collected through decision support services before the patient visit should offer much more substantive and valuable data to help the surgeon engage in more targeted and personalized conversations with the patient.

Phase One: Phase One identified several lessons about engaging clinicians in the process of shared decision making. When clinicians perceive a personal benefit in the decision support process, they may go well beyond the original implementation and enhance their practice by offering more seamless coordination of care and support for decision making, incorporating each patient's values and goals for care. In addition, when clinicians can assimilate real data about their performance, and the impact of new interventions and processes on their patients, more robust conversations ensue with patients ultimately producing better clinical outcomes.

Phase Two: Phase Two revealed the challenges of integrating data and workflow derived from a home-grown electronic health record, into a new, externally developed electronic health record. The new product invalidated existing data collection systems, making transition of pre-existing functionality more difficult. While anticipating and remedying all potential challenges in maintaining functionality from one electronic health record to another may be impossible, minimizing future barriers to electronic health record implementation may be achieved by developing a multidisciplinary team, debriefing, process revisions, and a commitment to participating in cycles of debriefings, revisions, and evaluations.

FIGURE 9.3
Decision quality, decision process, decision aid acceptability

DECISION QUALITY	
Patients unsure about their choice pre-decision aid	23%
Patients unsure about their post-decision aid	15%
Patients who switched after watching decision aid (includes = 10% who were unsure pre-decision aid) (n=174)	18%
DECISION PROCESS AFTER WATCHING THE DECISION AID	
Felt they knew the options, benefits and the risks	94%
Were clear about what was important	90%
Felt they had enough support to make a decision	87%
Were sure about their choice	71%
DECISION AID ACCEPTABILITY	
Patients had high ratings for decision aid content	94%
Would recommend them to others making the same decision	95%

Ultimately, achieved successes would have been impossible without a patient and family advisor on the multidisciplinary team. For patients, the patient and family advisor provided an invaluable perspective, linking clinical practice directly to patient experience. Having the patient and family advisor at the table also validated new processes by incorporating the voice of the focal point of any clinical unit: the patient.

Our recommendations to organizations embarking on similar journeys:

- Deployment of decision aids in practice must be planned carefully and well in advance;
- Facilitated sessions are the best means to debrief and design revisions when ideal workflows are not maintained;
- A formalized process must evaluate context-mechanism-outcomes;
- The success of any new system has to incorporate practical considerations for local teams who have a working knowledge of patient triage from process inception to completion.

COMMENTARIES

Relational Coordination Theory: Jody Hoffer Gittell
The change process documented at DHMC focused on shared decision making in the context of orthopaedic surgery. At the starting point there was a well-functioning home-grown electronic system for informing care providers when engaging with the patient. The system collected customized information from the patient, enabling good conversations between providers and patients about decisions regarding (for example) surgery and post-surgical recovery. The documented change process was driven by outside forces: a system-wide implementation of a new electronic health record, displacing the existing system. The new system was not designed in a way that supported shared decision making, and much effort went into retrofitting the new system to try to replicate the functionality of the original home-grown system.

In a sense, this case is a cautionary tale regarding the potentially disruptive impacts of externally imposed *structural interventions* which risk undermining relational coordination and coproduction rather than fostering it. I have heard complaints from care providers that electronic health records adopted by their health care systems have had the unintended effect of reducing relational coordination rather than strengthening it, by failing to link information from multiple perspectives on the patient and the patient's journey, thus reinforcing silos rather than breaking them down.

This cautionary tale is one that electronic health record developers need to hear, given the powerful potential for IT to either strengthen or weaken relational coordination and coproduction. It is also a cautionary tale for health care IT departments to consider as they choose external vendors. One critical selection criterion is the ability of that system to link across care providers (and between care providers and the patient) to create visibility into the patient as a whole person, something accomplished by linking insights from multiple perspectives across the patient's journey. Otherwise the IT investment is counterproductive and likely to reduce quality, efficiency, and patient well-being, and even financial performance as health systems move toward bundled payments and accountable care.

Normalization Process Theory: Glyn Elwyn
The digitalization of medical practice, like the digitalization of other aspects of life, demands some standardization of process. This is particularly true when many components join to form the information stream that is crucial in the care of patients. Demographic data, test data, diagnostic data, and information

about complex changes in therapy need to be managed as they constantly change over time. The need to agree on what is the work of digital data in medicine, who does it, how does it get done, and how is it measured, is critical to modern clinical processes. However, as this case illustrates so well, the use of patient decision support is not as well understood and is not subject to the same degree of agreement or assumed integration into workflows. A homegrown information system at DHMC had achieved high levels of agreement about NPT constructs, at least in a few clinical domains. But when a new system called eD-H was introduced, these old agreements about the work done by patient decision support interventions were swept aside by a rigid, predetermined, and largely inflexible system insufficiently malleable to allow customization of a local need. Normalization of patient decision support had been clearly initiated at DHMC but was not stable enough to withstand the introduction of powerful new systems of information management that failed to consider the role of these innovations in clinical pathways.

Microsystems: Marjorie M. Godfrey

The DHMC team's story highlights essential elements to applied microsystem theory, or developing patient-centered care while considering purpose, professionals, process, and patterns. The Center for Shared Decision Making at Dartmouth-Hitchcock represents a type of microsystem called a "supporting" microsystem. The video decision aids originally designed in Phase One supported patient decision making work within the orthopaedics clinical microsystem. The leadership of Dr. James Weinstein, who supported staff development in improvement knowledge, is impressive when considering the characteristics of high-performing microsystems. (We note that staff development was for medical staff, not all staff.) *Phase Two* describes Orthopaedics' advancement toward multidisciplinary teamwork including the patient and family advisor in the introduction of the electronic health record technology. The new technology forced the convening of a group charged with mapping current workflow and creating and testing the future resilience of a new system. High-performing microsystems enjoy regular meetings of such multidisciplinary groups to enhance knowledge of workflows, to discover opportunities for improvement and to realize interdependencies. The technology served as an intervention helping align the microsystem within the larger mesosystem. The case study demonstrates the enormous role of technology and information in engaging patients in shared decision making, thereby optimizing the care process and informing staff to enable better patient care design. Convening the multi-disciplinary team, studying workflow processes, and redesigning effective and efficient patient-centered systems to best meet patient and staff needs—all this enhanced the vision of providing the right patient with the right clinician at the right time.

The Spine Center's success and the creation of the Center for Shared Decision Making reflect the long history of patient-centeredness at Dartmouth-Hitchcock. Within DHMC, the Orthopaedics department continued the Spine Center's feedback process and worked to develop means of distributing decision aids to patients with knee osteoarthritis. A first step toward initial integration of decision aids into the knee clinic was mapping the microsystem's workflow, a time-consuming process that effectively allowed the DHMC team to discern the most accessible location for integrating decision aids into usual practice. Initial workflow required a referral to receive the decision aid following a meeting with a specialist. The Phase One process from 2005 to 2010 was fairly useful to get decision aids to patients. However, when the hospital (macrosystem) implemented EPIC the microsystem processes had to be reconfigured. What stands out from the DHMC story is that the ultimate electronic health record success came through the inclusion of a patient and family advisor who sat with the multidisciplinary planning team to build a new and manageable system when the new electronic health record disrupted care processes. This case suggests that knowledge of high-performing microsystem characteristics (e.g. regular multidisciplinary meetings including patients and families focused on integrating technology workflows redesign and improvement) together with a medical center's commitment to change can ultimately refashion a home-grown system of decision aid referral.

Reflection on the Commentaries
This case reports on a project designed as feasibility research and intended as a sustainable innovation beyond the term of the research. As each commentary makes clear, the design for sustainability of a successful and clinically meaningful innovation assumes the continued existence of current technologies and support systems. We could not have anticipated the purchase of EPIC, nor the juggernaut of inflexibility it would exemplify. The tough lessons learned during the adoption of a new electronic medical record that shapes practice more than it is shaped by practice are worth passing on to both developers and health systems deserve wider recognition. Innovation will get harder to accomplish.

Kate Clay

Please direct permission requests for Chapter 9 Kate Clay at catharine.f.clay@dartmouth.edu

References

1. Rycroft-Malone J, Kitson AL, Harvey G, McCormack B, Seers K, Titchen A, Estabrooks CA. Ingredients for change: Revisiting a conceptual framework. *Quality & Safety in Health Care.* 2002;11(2):174–180.

2. Ward C, McCormack B. Creating an adult learning culture through practice development. *Nurse Education Today.* 2000;20(4):259–266.

3. Wennberg JE. *Tracking Medicine: A researcher's quest to understand healthcare.* New York: Oxford University Press; 2010.

SECTION THREE
NEXT STEPS

CHAPTER 10

Where Are We? Where Do We Go Next?

Glyn Elwyn

> *The slave doctor prescribes what mere experience suggests, as if he had exact knowledge; and when he has given his orders, like a tyrant, he rushes off with equal assurance to some other servant...But the other doctor, who is a freeman, attends and practices upon freemen; and he carries his enquires far back, and goes into the nature of the disorder, he enters into discourse with the patient and with his friends, and is at once getting information from the sick man, and also instructing him as far as he is able, and he will not prescribe for him until he has first convinced him; at last...he attempts to effect a cure. Now which is the better way of proceeding? Is he the better who accomplishes his ends in a double way, or he who works in one way...?*
>
> Plato, Laws IV (360 B.C.)

The call for clinical practice to be "patient-centered" came in the 1980s[1], founded on the biopsychosocial model[2]. But the view that good clinicians pay more attention to individual needs goes back as far as we can find written thoughts about medical practice. So do the problems, as Plato notes here.

These seven case studies are part of a tradition of increasing the degree to which patients are informed and engaged in care processes. It needs emphasizing, however, that although the declared goal of most patient decision support tools developed to date is to act as an "adjunct" to the clinical process, there is little evidence that tools designed for patient use before or after clinical encounters have the desired impact on the dialogue that subsequently occurs in face-to-face meetings. This issue needs to be addressed by more research. Almost all of these case studies assume that using these tools, if only they are given to the right patient, will lead to more, and better, shared decision making in clinical interactions, an assumption that is largely untested. Some work does examine the use of tools designed for incorporation into clinical encounters[3-5], but much more is needed to consider the impact and sustainability of using briefer tools in the interaction.

The case studies paint a picture of repeated challenges to implementation across many different health care settings in both the US and the UK. The conclusion must be that it is difficult to introduce these tools into routine practice. A conceptual review using the normalization process model attempted to explain some of the reasons[6]. Caldon et al.[7] noted resistance to these tools by professionals, and a recent in-depth interview study about efforts to introduce web-based tools to the NHS illustrated professional resistance in referring these tools to patients[8]. In 2006, Gravel[9] described clinicians' reluctance to use patient decision support interventions (DESIs) because they considered them to be inapplicable to their patients and clinical situations. Légaré et al.[10] more recently reviewed 6,764 titles and abstracts and analyzed five random control trials, tentatively concluding that promoting shared decision making may depend on (a) training health care professionals and (b) adopting patient-targeted DESIs[10]. So although two decades of research have established the positive effect of using well-designed patient DESIs—that patients gain knowledge, greater understanding of probabilities, and increased confidence in decisions[11]—the challenge remains of sustaining their use in routine care, and these case studies, from leading researchers in the field, confirm the problem.

Nevertheless, the policy context has gradually improved in recent years. In the US, the 2010 Affordable Care Act[12] explicitly promotes shared decision making and use of DESIs. Some states have passed legislation supporting their use[13]. Similarly, in the UK, shared decision making has been at the center of policy developments[14] with investments made in developing online DESIs[15]. The Canadian province of Saskatchewan now supports province-wide work in the use of DESIs[16]. Many other countries are alert to the benefits, and are considering policy developments in this area[17].

Yet despite these policy developments and the existence of controlled trials demonstrating the efficacy of these interventions[11], their effectiveness and adoption into mainstream clinical practice is not yet established. Many reports of early implementation efforts in the field are not yet published in the peer-reviewed literature[18]. For over a decade, the Dartmouth-Hitchcock Medical Center in Lebanon, New Hampshire, has routinely provided many patients with DESIs (DVDs and booklets) through their Center for Shared Decision Making[19]. Group Health in Seattle has reported organization-wide adoption of DESIs for selected conditions[20,21]. However, these remain relatively isolated examples of adoption. Many researchers develop and evaluate these tools work in academic settings, but no studies of sustained wide-scale adoption have yet been reported.

Recently, in 2010, the International Patient Decision Aids (IPDAS) Collaboration initiated a review of its quality dimensions to inform recommendations about how best to implement patient DESIs into practice, thus reflecting the increasing emphasis on delivery research encompassing implementation or improvement science[22]. Pronovost & Goeschel highlight an issue of central importance for policy-makers—examining why interventions that have positive effects for patients under controlled conditions do not become established in routine settings[22]. To address this gap in knowledge, a group searched for and analyzed the findings of peer-reviewed studies that investigated the effectiveness of strategies, methods, or approaches to *implement* patient-targeted DESIs into routine clinical settings and workflows.

The detailed report of their findings is published elsewhere[23]: a summary appears here. Despite increasing interest in moving patient decision support from the world of randomized trials to routine settings, major implementation challenges existed. In contrast to the positive findings reported in trials[11], the review illustrated widespread professional indifference and organizational inertia. Many barriers resemble those encountered in other attempts to improve practice performance, where competing priorities fight for precedence, and uncertainty about the added value of the proposed intervention favors the status quo[24]. The organizations included in the studies were willing volunteers and so implementation might be even more difficult in other settings. Although many countries are considering adopting shared decision making, most implementation work to date has been located in North America. Ten studies came from the US and three from Canada, an illustration that this work remains in a tightly circumscribed research domain[17]. Most of the work was conducted with limited resources in comparison to research funded by mainstream sources, such as the National Institutes of Health, and so in appraising these studies we need to recognize these constraints.

The studies did reveal issues specific to the challenge of implementing patient DESIs. Relying on clinicians to refer patients to these tools leads to limited utilization, and so system-based approaches, where feasible, will reach more patients. Unfortunately, system approaches rely on identifying eligible patients ahead of clinical encounters, a task that is possible for only a limited number of conditions. Even when this is feasible, logistical and infrastructure challenges often impede integration of these interventions into practice. When patients present with undifferentiated problems, identifying their decision support needs ahead of a visit may be impossible. This issue limits the scope for studies adopting a referral model: most are based on clinical issues where prior identification is possible, e.g. invitations for screening and prevention. Yet even in secondary care where it is often possible to predict the scope of clinical decisions, the process of ensuring that patients use DESIs ahead of encounters poses challenges because the window of opportunity is often short. Ultimately, the studies indicatec that this degree of logistical infrastructure is hard to initiate and maintain, and will require sustained investment[25-30]. Referral by clinicians at the point of care will continue to be necessary for many clinical issues for which decision support is available.

The included studies used a "referral model" of DESI dissemination whereby practitioners or their support staff identified patients as eligible for decision support. The referral model proposes that these tools are "adjuncts" that support shared decision making, when used ahead of visits or shortly afterwards[31]. However, the claim that these tools are positively viewed as "adjuncts" by clinicians seems unsupported in practice. Many of the studies report that professionals distrust the content of the tools, question their evidence-base, believe that they do not reflect "local" data, think that patients will decline to take part in decisions, and, critically, that offering options is not what they would advocate from a "best practice" perspective. These findings suggest that the reluctance to prioritize DESIs might lie deeper than a general resistance to change. The referral model might be based on assumptions about their contribution that is not shared by front-line clinicians[7], a suggestion we discuss further below. An alternative model whereby the practitioner initiates shared decision making in the space of clinical encounters, using briefer DESIs to catalyze dialogue about options, which in turn could lead to the use of more extensive tools[32], does not seem to have been extensively investigated, although a few trials exist[33,34].

Although studies described many barriers to implementation (see Table 10.1), these barriers were seldom examined in depth, with the exception of three studies that employed qualitative interviews[28,35,36]. Additional insights might have been gained if more studies had explored the views of professionals regarding the use of DESIs and specifically about their impact on practice workflows.

TABLE 10.1
Barriers and facilitators to implementation

	BARRIERS	**FACILITATORS**
INTERPERSONAL	• Provider attitude toward and understanding of how to use DESIs and engage in SDM. • Provider view that patients don't want to be responsible for making decisions. • Provider belief that DESIs were in 'competition' with other information designed for patients. • Provider view that distributing DESIs was not part of their role or responsibility.	• Training and skills development for providers on how to use DESIs and engage in SDM. • Identification of a clinical champion.
SYSTEMIC	• Competing clinical demands. • Time pressure or lack of time. • Difficulty identifying eligible patients. • Lack of organizational support.	• Automated identification of eligible patients. • Patient use of DESIs prior to relevant clinical consultations.

The challenge of implementing patient DESIs is already well documented[9,10] and we also know that practitioners often don't achieve shared decision making[37]. We must be careful not to equate the successful introduction of DESIs into clinical pathways as automatically leading to shared decision making. For instance, Frosch et al. found that the use of a prostate specific antigen DESI ahead of a clinical encounter led to less shared decision making if the patient didn't favor screening[38]. While at the patient level these interventions have positive results[11], we do not yet fully understand their impact on clinician-patient dialogue. Other models where practitioners use brief tools and take more responsibility for initiating shared decision making face-to-face with patients deserve further investigation.

Similarly, more use could be made of developments in how to evaluate complex interventions[39], implementation, and evaluation studies[40], to determine why interventions that have a positive effect in some settings, fail in other settings: or, that context matters[41]. Many opportunities exist to bring these worlds of inquiry to bear on how best to implement patient DESIs. Damschroder et al have provided a consolidated framework for advancing implementation science[40], a synthesis of 19 models that describes five domains, namely, intervention characteristics, outer setting, inner setting, characteristics of individuals involved, and the process of implementation. Future studies should consider the reported utility of these conceptual frameworks to guide implementation.

The review concludes by stating that it seems too early for making recommendations about how best to introduce patient decision support into routine practice. Although it would be easy for us to suggest general principles of successful adoption[42], the reviewing authors concluded that the first step is to elucidate more clearly the specific underlying issues that militate against the use of patient DESIs—such as incentive and performance frameworks that reward behaviors at odds with informing and involving patients. These frameworks are important contextual explanations for the difficulties experienced in implementation.

Where do we go next?

The first step is to be clear about goals. Much of the work we see here is about trying to implement a solution that works around the real problem—that health professionals do not place much effort into, nor do they value, the process of informing and involving patients in health care decisions. Here, is a list of items to be tackled:

- Recognize that there is confusion about terminology and core concepts, especially at the level of policy setting.
- Consider that shared decision making might not be the best name for different audiences, despite the attention received at policy levels.
- Recognize that clinicians are key to this area of work—they have decisional power, at least as perceived by most patients.
- Recognize that patients are often afraid of asking questions, stating their preferences, and disagreeing; not all patients, but more than we think, and that this is especially true if patients are disadvantaged or from cultures where questioning authority runs counter to social norms.
- Recognize that existing financial systems are at cross-purposes—that professional cultures and incentive frameworks reward behaviors that are often at odds to achieving the informed patient preference.

- Recognize that until we can measure more effectively what we mean by involving patients, we will be unable to make progress.
- Recognize that, until patients demand change, the stakeholders that need to be engaged are clinicians and that to make progress we must address their dominant paradigm of unquestioned expertise by asking them to rise to the challenge of diagnosing preferences as well as diagnosing disease.
- Finally, undertake research that addresses real world problems of implementation and make sure the work is better reported.

Recognize that there is confusion about terminology and core concepts
Put simply, we cannot conflate the work of "sharing decisions" with the work of trying to get patient DESIs into the hands of patients. No matter how many trials of these tools achieve positive short-term or even medium-term outcomes, the impact on practice in the real world is close to zero unless:

1. We can be confident that the same impact can be demonstrated in real-world settings and
2. These tools can be embedded into clinical workflows and truly sustained.

Consider that "shared decision making" might not be the best name
Maybe we need to consider that using the term "shared decision making" does not help the effort to spread the core aspects of patient-centered care. The term was a wonderful shorthand for building a community of researchers around the idea of skills and tools to help achieve the goal of informing and involving patients. But if the goal now is to spread this idea across the globe, to all patients and to all clinicians, perhaps the term has outlived its usefulness. Clinicians, we might as well face it, rather like to be regarded as experts and are slow to adopt a more egalitarian stance, however much we hope that this occurs. In any case, we also often hear that they already "share" decisions with patients—so they see no problem that they have to solve. Shared decision making to doctors is either a threat to their status or an insult to their existing performance. Neither is a good reaction when the goal is to foster curiosity.

From the patient perspective, there is also a difficulty with the term. As many have noted, many patients, when asked, do not consider themselves as health care decision makers. So the very idea that they would be responsible for making decisions is new to them, and their reaction is either to worry about the responsibility that this confers or to anticipate regret, particularly if they make what might turn out to be a "wrong" decision. This anxiety about decisional responsibility is heard in the question often put to health care professionals: "What would you do if you were me?" As multiple surveys have shown, patients want to be informed, they want their views elicited and heard, but as for being responsible for the ultimate decision—that, for many patients, is a step too far. So the term "shared decision making" conjures up, at best, the burden of unfamiliar extra work, or, at worst, the perception of professionals abrogating their role as expert guides.

Recognize that clinicians have decisional power, at least as perceived by patients

Undoubtedly patients are becoming more informed, more active, and less accepting of professionals who do not explain, are not open to questions, and unwilling to pay respect to their individual needs and wants. The information age will accelerate this trend, and social media will empower many patients to become very knowledgeable. Some patients, especially those with chronic illness, will overtake the expertise of their clinicians. They will be the pioneers of a profound change in role. But many patients, even those with undoubted intellectual capacity to become informed (particularly when ill), will actively choose to subject themselves to the guidance of a trusted professional. They know they could behave differently and become exquisitely well-informed, but they choose to be passive. In such situations, patients often want professionals to understand what matters to them, and then make good recommendations. Many patients choose to leave the decisional responsibility in the hands of their professional. We cannot work against this tendency. There is often a sense that those who advocate for "shared decision making" have not accepted that patients actively choose to decline decisional responsibility.

Recognize that patients often fear asking questions

Oddly enough very few studies have asked patients why they are passive in asking questions, stating preferences, and disagreeing with health professionals. It is as if the researchers have assumed that patients, when given more information, would make good use of their new knowledge and change their behavior, adopting new and more active roles.

It seems that we could not be more mistaken. Even patients who are wealthy, educated, and live in the most privileged locations dare not challenge their health professionals lest they be viewed as difficult or awkward[43]. They do this not because they want to remain polite and not because they lack questions. They do this because they do not want to risk being viewed as "bad patients" undeserving of high-quality care. They worry that by asking questions and voicing preferences they would end up receiving worse care. What chance, then, for patients from disadvantaged brackets of society or from cultures where questioning authority is counter to social norms? The gradient of the medical power hierarchy is steeper and more slippery than we realized.

Recognize that many existing financial systems are at cross-purposes

The best analogy to how health care operates, especially in the US, is the hotel industry. If your patient is in the hospital, financial incentives align to encourage doing as many tests as possible, as many procedures as are deemed needed, and for the patient to stay in a bed as long as he or she wishes. The insurance is paying and the providers make a profit on a fee-for-service basis. Hotels work in the same way, only, normally, you are the one paying. Similarly, systems also pay for performance and efficiency: see as many patients as possible and as quickly as possible; make sure that targets are met, and more recently, that new clinical performance targets have been set, (e.g. ensure that patients are encouraged to use medication to reach biomedical targets such as agreed blood pressure levels or low cholesterol levels).

These incentives and performance measures do not reward health professionals for taking time to explain, to compare treatment options, or to make sure patients understand the harms as well as the benefits of treatments. In fact, the new systems punish the professionals or organizations working on a fee-for-service basis. Well-informed patients, on average, choose less health care. This equates to less profit. That is not good news for the ways things are currently designed.

Recognize we will make no progress until we can measure

We can measure shared decision making or at least we can measure the key behaviors that need to be exhibited by a handful of existing measures[44–46]. This is laborious and expensive work. Audio or video tapes need to be collected, observed, and assessed. It is the domain of academic research.

A better way is asking patients whether or not they have experienced "shared decision making" and whether they have made a "high-quality" decision. Despite many efforts over the last decade, this goal has proved elusive. Patients give high scores regardless of objective evidence on efforts made to inform and involve them in decisions. Many scales have tried to tackle this challenge and all report the same difficulties[47]. Progress will depend on the development of valid measurement methods, and even better if they can be used in routine care as a means of providing feedback to providers and systems.

Recognize clinicians as the key route to change

Major strides have been made in the design of complex and elaborate decision support tools for patients and the trial evidence is convincing: they are indeed effective when patients are asked to use them[11]. But evidence suggests that it is incredibly difficult to get these tools into routine care, and furthermore we do not know much about what happens in the "black box" (the clinical encounter) when these tools are used. Perhaps it is time to recognize the central role of clinicians: that until clinicians become motivated to inform and involve patients, we will observe very little change. One recent attempt to make progress has been to pose the work as a challenge that can be met by good and expert clinicians and that to diagnose preferences as well as diagnosing disease is a pinnacle of effective practice[48]. Different types of tools will be needed, of course: tools that can be used collaboratively by both patients and professionals[49].

Research the real-world problems of implementation

There are well over 80 trials of decision support tools, most designed for use outside the clinical encounter. An argument could be made that no further trial of these tools is required—that the next frontier is how to implement these tools into clinical practice. We should pause, however, before accepting this argument. It might be well worth asking another, more fundamental question. Are we addressing the right problem? Perhaps the questions need go deeper? Might it be that we should investigate why professionals behave the way they do? What drives so many to spend so little time explaining treatments to patients? Why so little curiosity about individual preferences? Professionals do not go to work to do a bad job, so what is it about the surrounding systems that make it so difficult to become patient-centered?

To answer these types of questions, we might need approaches not previously adopted by this field. How should new delivery systems be implementing change[50]? For instance, methods such as cognitive task analysis, ethnography and action research, tools to assess the "adaptive reserve" of teams[51] or their "readiness for change"[52], are all approaches that would pay more attention to the role of the key stakeholders in shaping and using the technologies[53], and how they fit into the demands of other technologies, such as the electronic medical record and demands for performance metrics. Amidst all of this is the need to monitor which professional and team-related behaviors will be rewarded as health systems increasingly seek to ensure that patients experience better quality of care[54].

---

*Please direct permission requests for Chapter 10 to Glyn Elwyn
at glynelwyn@gmail.com*

References

1. Levenstein JH. The patient-centred general practice consultation. *South African Family Practice*. 1984;5(9):276–282.

2. Engel GL. The need for a new medical model: A challenge for biomedicine. *Science (New York, N.Y.)*. 1977;196(4286):129–136.

3. Elwyn G, Edwards AG, Hood K, Robling MR, Atwell C, Russell I, Wensing M, Grol R. Achieving involvement: Process outcomes from a cluster randomized trial of shared decision making skill development and use of risk communication aids in general practice. *Family Practice*. 2004;21(4):337–346.

4. Nannenga MR, Montori VM, Weymiller AJ, Smith SA, Christianson TJH, Bryant SC, Gafni A, Charles C, Mullan RJ, Jones LA, Bolona ER, Guyatt GH. A treatment decision aid may increase patient trust in the diabetes specialist. The Statin Choice randomized trial. *Health Expectations*. 2009;12(1):38–44.

5. Whelan T, Gafni A, Charles C, Levine M. Lessons learned from the Decision Board: A unique and evolving decision aid. *Health Expectations*. 2000;3(1):69–76.

6. Elwyn G, Légaré F, Van der Weijden T, Edwards AG, May CR. Arduous implementation: Does the Normalisation Process Model explain why it's so difficult to embed decision support technologies for patients in routine clinical practice. *Implementation Science*. 2008;3(1):57.

7. Caldon LJM, Collins KA, Reed MW, Sivell S, Austoker J, Clements AM, Patnick J, Elwyn G, BresDex Group. Clinicians' concerns about decision support interventions for patients facing breast cancer surgery options: Understanding the challenge of implementing shared decision-making. *Health Expectations*. 2011;14(2):133–146.

8. Elwyn G, Rix A, Holt T, Jones D. Why do clinicians not refer patients to online decision support tools? Interviews with front line clinics in the NHS. *BMJ Open*. 2012;2(6).

9. Gravel K, Légaré F, Graham ID. Barriers and facilitators to implementing shared decision-making in clinical practice: A systematic review of health professionals' perceptions. *Implementation Science*. 2006;1:16.

10. Légaré F, Ratté S, Stacey D, Kryworuchko J, Gravel K, Graham ID, Turcotte S. Interventions for improving the adoption of shared decision making by healthcare professionals. *Cochrane Database of Systematic Reviews*. 2010;5(5).

11. Stacey D, Bennett CL, Barry MJ, Col NF, Eden KB, Holmes-Rovner M, Llewellyn-Thomas HA, Lyddiatt A, Légaré F, Thomson RG. Decision aids for people facing health treatment or screening decisions. *Cochrane Database of Systematic Reviews*. 2011;10(10).

12. Senate and House of Representatives. *The Patient Protection and Affordable Care Act.* Washington: 111th Congress, 2nd Session; 2010.

13. Washington State Legislature. *Shared decision-making demonstration project--Preference-sensitive care.* RCW 41.05.033; 2007.

14. Department of Health. *Equity and excellence: Liberating the NHS.* London: The Stationery Office Limited on behalf of the Controller of Her Majesty's Stationery Office; 2010.

15. Elwyn G, Laitner S, Coulter A, Walker E, Watson P, Thomson RG. Implementing shared decision making in the NHS. *BMJ.* 2010;341(c5146).

16. Légaré F, Stacey D, Forest P-G, Coutu M-F. Moving SDM forward in Canada: Milestones, public involvement, and barriers that remain. *Zeitschrift für Evidenz, Fortbildung und Qualität im Gesundheitswesen.* 2011;105(4):245–253.

17. Härter M, Van der Weijden T, Elwyn G. Policy and practice developments in the implementation of shared decision making: An international perspective. *Zeitschrift für Evidenz, Fortbildung und Qualität im Gesundheitswesen.* 2011;105(4):229–233.

18. Informed Medical Decisions Foundation. *Research and Implementation Report.* Boston; 2011.

19. The Dartmouth Institute for Health Policy and Clinical Practice. *Improving value-based care and outcomes of clinical populations in an electronic health record system environment.* Hanover, New Hampshire; 2011.

20. Arterburn D, Wellman R, Westbrook EO, Rutter C, Ross TR, McCulloch D, Handley M, Jung C. Introducing decision aids at Group Health was linked to sharply lower hip and knee surgery rates and costs. *Health Affairs.* 2012;31(9):2094–2104.

21. Hsu C, Liss DT, Westbrook EO, Arterburn D. Incorporating patient decision aids into standard clinical practice in an integrated delivery system. *Medical Decision Making.* 2013;33(1):85–97.

22. Pronovost PJ, Goeschel CA. Time to take health delivery research seriously. *JAMA.* 2011;306(3):310–311.

23. Elwyn G, Scholl I, Tietbohl C, Mann M, Edwards AG, Clay C, Légaré F, Van der Weijden T, Lewis CL, Wexler RM, Frosch DL. Many miles to go...: A systematic review of the implementation of patient decision support interventions into routine clinical practice. *BMC Medical Informatics and Decision Making.* 2013;In Press.

24. Cabana MD, Rand CS, Powe NR, Wu AW, Wilson MH, Abboud P-AAC, Rubin HR. Why don't physicians follow clinical practice guidelines? A framework for improvement. *JAMA.* 1999;282(15):1458–1465.

25. Belkora JK, Loth MK, Volz S, Rugo HS. Implementing decision and communication aids to facilitate patient-centered care in breast cancer: A case study. *Patient Education and Counseling.* 2009;77(3):360–368.

26. Belkora JK, Teng A, Volz S, Loth MK, Esserman LJ. Expanding the reach of decision and communication aids in a breast care center: A quality improvement study. *Patient Education and Counseling.* 2011;83(2):234–239.

27. Feibelmann S, Yang TS, Uzogara EE, Sepucha KR. What does it take to have sustained use of decision aids? A programme evaluation for the Breast Cancer Initiative. *Health Expectations.* 2011;14 Suppl 1:85–95.

28. Frosch DL, Singer KJ, Timmermans S. Conducting implementation research in community-based primary care: A qualitative study on integrating patient decision support interventions for cancer screening into routine practice. *Health Expectations.* 2011;14 Suppl 1:73–84.

29. Miller KM, Brenner AT, Griffith JM, Pignone MP, Lewis CL. Promoting decision aid use in primary care using a staff member for delivery. *Patient Education and Counseling.* 2012;86(2):189–194.

30. Silvia KA, Sepucha KR. Decision aids in routine practice: Lessons from the breast cancer initiative. *Health Expectations.* 2006;9(3):255–264.

31. O'Connor AM, Wennberg JE, Légaré F, Llewellyn-Thomas HA, Moulton BW, Sepucha KR, Sodano AG, King JS. Toward the "tipping point": Decision aids and informed patient choice. *Health Affairs.* 2007;26(3):716–725.

32. Elwyn G, Frosch DL, Thomson RG, Joseph-Williams N, Lloyd A, Kinnersley P, Cording E, Tomson D, Dodd C, Rollnick S, Edwards AG, Barry MJ. Shared decision making: A model for clinical practice. *Journal of General Internal Medicine.* 2012;27(10):1361–1367.

33. Edwards AG, Elwyn G, Hood K, Atwell C, Robling MR, Houston H, Kinnersley P, Russell I. Patient-based outcome results from a cluster randomized trial of shared decision making skill development and use of risk communication aids in general practice. *Family Practice.* 2004;21(4):347–354.

34. Mann DM, Ponieman D, Montori VM, Arciniega J, McGinn T. The Statin Choice decision aid in primary care: A randomized trial. *Patient Education and Counseling.* 2010;80(1):138–140.

35. Stapleton H, Kirkham M, Thomas G. Qualitative study of evidence based leaflets in maternity care. *BMJ.* 2002;324(7338):639.

36. Uy V, May SG, Tietbohl C, Frosch DL. Barriers and facilitators to routine distribution of patient decision support interventions: A preliminary study in community-based primary care settings. *Health Expectations.* 2012;Epub.

37. Couët N, Desroches S, Robitaille H, Vaillancourt H, Turcotte S, Elwyn G, Légaré F. Using OPTION to assess the level to which health professionals involve patients in decision-making: A systematic review. *In review.* 2012.

38. Frosch DL, Légaré F, Mangione CM. Using decision aids in community-based primary care: An evaluation with ethnically diverse patients. *Patient Education and Counseling.* 2008;73(3):490–796.

39. Campbell NC, Murray E, Darbyshire J, Emery J, Farmer A, Griffiths F, Guthrie B, Lester H, Wilson P, Kinmonth AL. Designing and evaluating complex interventions to improve health care. *BMJ.* 2007;334(7591):455–459.

40. Damschroder LJ, Aron DC, Keith RE, Kirsh SR, Alexander JA, Lowery JC. Fostering implementation of health services research findings into practice: A consolidated framework for advancing implementation science. *Implementation Science.* 2009;4:50.

41. Pawson R, Tilley N. *Realistic Evaluation.* London: Sage Publications Ltd; 1997.

42. Greenhalgh T, Robert G, Macfarlane F, Bate P, Kyriakidou O. Diffusion of innovations in service organizations: Systematic review and recommendations. *The Milbank Quarterly.* 2004;82(4):581–629.

43. Frosch DL, May SG, Rendle KAS, Tietbohl C, Elwyn G. Authoritarian physicians and patients' fear of being labeled "difficult" among key obstacles to shared decision making. *Health Affairs.* 2012;31(5):1030–1038.

44. Braddock CH, Edwards KA, Hasenberg NM, Laidley TL, Levinson W. Informed decision making in outpatient practice: Time to get back to basics. *JAMA.* 1999;282(24):2313–2320.

45. Elwyn G, Hutchings H, Edwards AG, Rapport F, Wensing M, Cheung W, Grol R. The OPTION scale: Measuring the extent that clinicians involve patients in decision-making tasks. *Health Expectations.* 2005;8(1):34–42.

46. Shields CG, Franks P, Fiscella K, Meldrum S, Epstein RM. Rochester participatory decision-making scale (RPAD): Reliability and validity. *Annals of Family Medicine.* 2005;3(5):436–442.

47. Scholl I, Koelewijn-van Loon M, Sepucha KR, Elwyn G, Légaré F, Härter M, Dirmaier J. Measurement of shared decision making - a review of instruments. *Zeitschrift für Evidenz, Fortbildung und Qualität im Gesundheitswesen.* 2011;105(4):313–324.

48. Mulley AG, Trimble C, Elwyn G. Stop the silent misdiagnosis: Patients' preferences matter. *BMJ.* 2012;345:e6572.

49. Elwyn G, Lloyd A, Joseph-Williams N, Cording E, Thomson RG, Durand M-A, Edwards AG. Option Grids: Shared decision making made easier. *Patient Education and Counseling.* 2013;90(2):207–212.

50. Alexander JA, Hearld LR. Methods and metrics challenges of delivery-system research. *Implementation Science.* 2012;7(1):15.

51. Nutting PA, Crabtree BF, Miller WL, Stange KC, Stewart E, Jaén C. Transforming physician practices to patient-centered medical homes: Lessons from the national demonstration project. *Health Affairs.* 2011;30(3):439–445.

52. Weiner BJ, Amick H, Lee S-YD. Conceptualization and measurement of organizational readiness for change: A review of the literature in health services research and other fields. Medical Care Research and Review. 2008;65(4):379–436.

53. Greenhalgh T, Wieringa S. Is it time to drop the "knowledge translation" metaphor? A critical literature review. *Journal of the Royal Society of Medicine.* 2011;104(12):501–509.

54. National Institute for Health and Clinical Excellence. *Quality standard for patient experience in adult NHS services.*; 2012.

APPENDIX I

List of Authors and Editors

David Arterburn is a general internist and health services researcher with the Group Health Research Institute in Seattle. His research covers a broad range including shared decision making related to elective surgery, weight management interventions, pharmacoepidemiology, and bariatric surgery. He serves as a medical editor for the Informed Medical Decisions Foundation.

Alison Brenner is a doctoral candidate at the University of Washington, Seattle, with an interest in patient medical decision making and patient education implementation methods.

Kate Clay is a nurse whose expertise in decision support has led to her current position as director of education and outreach at The Dartmouth Institute for Health Policy and Clinical Practice (TDI), focused on teaching distance-learning courses on shared decision making. Affiliations: The Geisel School of Medicine at Dartmouth; Dartmouth-Hitchcock Medical Center.

Cristin Colford is a clinician-educator and general internist interested in practicing shared decision making with patients as well as teaching this skill to medical students and residents. Affiliations: Division of General Medicine, University of North Carolina School of Medicine.

Carol DuBois, formerly a financial analyst and business owner, served as a patient advisor on multiple assignments over the past five years including working on "My Quest," which addressed patient questionnaires in Dartmouth-Hitchcock Medical Center's transition to the EPIC system. A longtime Hanover resident, she has been an active participant in the community including twenty years as a docent at the Hood Museum of Art.

Glyn Elwyn is a clinician-researcher with an interest in shared decision making, user-centered design of decision support tools, and their integration into routine health care. Affiliations: Cardiff University; Dartmouth College; Radboud University, Nijmegen, Netherlands.

Dominick L. Frosch is a clinical health psychologist interested in shared decision making and design, evaluation, and implementation of patient decision support in routine clinical care. Affiliations: Palo Alto Medical Foundation Research Institute; Department of Medicine, University of California, and the Gordon and Betty Moore Foundation.

Jody Hoffer Gittell is an organizational scholar with interests in organizational effectiveness and relational forms of coordination, recently extending beyond coordination to include coproduction with patients, families and the broader community. Affiliations: Professor of Management, Heller School at Brandeis University; Executive Director, Relational Coordination Research Collaborative.

Marjorie M. Godfrey is an instructor at The Dartmouth Institute for Health Policy and Clinical Practice. She is also co-director of The Dartmouth Institute's Microsystems Academy and director of the Clinical Microsystems Resource Group. Marjorie is a national leader of designing and implementing improvement strategies targeting the place where patients, family, and care teams meet the clinical microsystem.

Stuart W. Grande is a post-doctoral fellow at The Dartmouth Center for Health Care Delivery Science (TDC) focusing on implementing shared decision making practice and improving patient-provider communication.

Clarissa Hsu is a medical anthropologist with expertise in conducting research and evaluation on a wide range of issues including clinical quality improvement, patient-centered care, complementary and alternative medicine, and community-based health improvement. She is a faculty member at Group Health Research Institute's Center for Community Health and Evaluation in Seattle.

Carmen L. Lewis is a practicing general internist and a health communication researcher with expertise in cancer screening and prevention and medical decision making at the University of North Carolina.

David T. Liss is a health services researcher interested in evaluating new models of care delivery for patients with chronic conditions, such as the patient-centered medical home. Affiliations: University of Washington Department of Health Services; Group Health Research Institute, Seattle.

Robert Malone is a clinical pharmacist-practitioner with experience in quality improvement and design, implementation, and evaluation of planned-care programs for vulnerable patients, particularly those with diabetes. He is associate professor of medicine and pharmacy and vice president of Health Care Practice Quality and Innovation at the University of North Carolina.

Melanie P. Mastanduno is managing director for a project at The Dartmouth Institute for Health Policy and Clinical Practice (TDI) that tracks patient-reported measures of functional status and health risk as key outcomes of care. During the previous decade, she directed the launch of "Quality Reports," a nationally recognized website for information on clinical outcomes, patient satisfaction, and service charges. She holds a professional nursing degree and an MPH from the Johns Hopkins University School of Hygiene and Public Health.

Suepattra G. May is an assistant research anthropologist in the department of Health Services Research at the Palo Alto Medical Foundation Research Institute. Dr. May received her doctorate in medical anthropology at the University of California, Berkeley and San Francisco, and her MPH in health services from the University of Washington.

Shaun McDonald is a business intelligence analyst with the University of North Carolina Information Services supporting the Health Care System's Meaningful Use and quality improvement objectives.

Lawrence E. Morrissey Jr. is a general pediatrician and medical director of quality improvement for Stillwater Medical Group (Minnesota) and has been working on implementation of shared decision making since 2006.

Karen Sepucha is a researcher with an interest in shared decision making and decision aids, measurement of decision quality, and integration of shared decision-making into primary and specialty care. Affiliations: Massachusetts General Hospital; Harvard Medical School.

Leigh Simmons is a clinician and researcher with an interest in implementation of shared decision making in primary care practice and clinician training in shared decision making skills. Affiliations: Massachusetts General Hospital; Harvard Medical School.

Richard Thomson is associate dean for patient and public engagement at Newcastle University, England, leading the Decision Making and Organization of Care research program. He has led many studies in shared decision making, and co-leads two NIHR program grants: Shared Decision Making and hyperacute stroke, and patient involvement in safety.

Caroline Tietbohl is a research assistant with an interest in shared decision making, implementation of decision support tools and their subsequent impact on routine health care. Affiliation: Palo Alto Medical Foundation Research Institute.

Ivan Tomek is an orthopaedic surgeon whose hip and knee arthritis practice incorporates shared decision making; he conducts clinical research related to patient-centered care and comparative clinical effectiveness. Affiliations: The Geisel School of Medicine; Dartmouth-Hitchcock Medical Center; The Dartmouth Institute for Health Policy and Clinical Practice.

Dale Collins Vidal is a professor of surgery at The Geisel School of Medicine at Dartmouth and Chief of Plastic Surgery at Dartmouth-Hitchcock Medical Center. As a leader in health care transparency and shared decision making, Dr. Vidal was named the Director for the Center for Informed Choice in October 2008.

Matthew Waters worked as a care assistant in the University of North Carolina internal medicine practice until July 2012. He is now a first-year medical student at the University of North Carolina Medical School.

Emily O. Westbrook is manager of the Research Project Management Office at Group Health Research Institute in Seattle, Washington. She has extensive experience in the development, implementation, and administration of grant-funded research in the areas of health behavior and medical decision making.

Kim Young-Wright is a quality improvement specialist and project coach who has worked in elder services and health care with a focus on process improvements to facilitate patient-centered treatment and optimal outcomes. Affiliation: University of North Carolina Healthcare.

APPENDIX II

Selected Glossary

Clinical Champion
An individual health provider who is a member of a clinical team and serves as a supportive voice for innovation within the clinical unit and who may be used as an ally in research efforts to implement novel workflow and process changes.

Clinical Microsystem
The small functional front-line units that provide most health care to most people. Microsystems are essential building blocks of larger organizations and of the health system, as the place where patients, families, and care teams meet. The quality and value of care produced by the large health system can be no better than the care and services generated by the small systems of which it is composed.

Concordance
This term refers to the alignment of patient values and the treatment prescribed.

Decision Aids
See patient decision support interventions.

Electronic Medical Records
According to the Institute of Medicine (IOM) the core functions of an electronic medical record should include the following: health information and data, results management, order management, decision support, electronic communication and connectivity, patient support, administrative processes, and reporting.

Fee for service
A type of payment model for health care providers: it means that providers are paid a certain amount based on services provided, e.g., knee surgeries, hip replacements, or heart stents. If a provider provides more services, the provider earns more money.

Patient and Family Advisor
A hospital representative who supports patients and families in the clinical setting to provide the most effective patient-centered experience possible. In many cases these are former patients who have had long-time experience with health care. Often the PFA serves as a voice for patients and families on hospital boards, care units, and national associations.

Patient-Centered Care
Health care that establishes a partnership among practitioners, patients, and their families (when appropriate) to ensure that decisions respect patients' wants, needs, and preferences, and that patients have the necessary education and support to make decisions and participate in their own care.

Patient Decision Support Interventions (DESIs)
Decision support interventions help people think about choices they face: they describe where and why choice exists; they provide information about options, including, where reasonable, the option of taking no action. These interventions help people to deliberate, independently or with others, about options by considering relevant attributes to help them forecast how they might feel about short, intermediate, and long-term outcomes which have relevant consequences; in ways which help support the process of constructing preferences and eventual decision making, appropriate to their individual situation.

Professional Equipoise
The clinician's portrayal that more than one reasonable way exists to manage a health issue, including the option of taking no action.

Preference-Sensitive Decisions
Preference-sensitive decisions are those where there is sufficient uncertainty about the most effective means of treatment, or where patient preferences can legitimately overrule the existence of good evidence of clinical effectiveness. Some might argue that all decisions faced by patients, are, by definition (or should be) sensitive to their preferences. In most legislatures, acting against patient preferences is considered assault.

Shared Decision Making
Shared decision making is an approach where clinicians and patients communicate using the best available evidence when faced with the task of making decisions, and where patients are supported to deliberate about the possible attributes and consequences of options, to arrive at informed preferences in determining the best course of action, where this is desired, ethical, and legal.

APPENDIX III

Institute Team

CO-CHAIRS

Dale Collins Vidal, MD, MS
Director, The Center for Informed Choice, The Dartmouth Institute
 for Health Policy and Clinical Practice
Medical Director, Center for Shared Decision Making
Chief, Plastic Surgery, Dartmouth-Hitchcock Medical Center
Professor of Surgery & Community & Family Medicine,
 Geisel School of Medicine,
Lebanon, NH, USA

Glyn Elwyn, BA, MB, BCh, MSc FRCGP, PhD
Professor of Primary Care Medicine
Research Director, Department of Primary Care and Public Health,
 Cardiff University
General Practitioner, North Cardiff Medical Centre
Cardiff, Wales, UK

MANAGEMENT TEAM

Allison J. Hawke, BA - Institute Manger
Associate Director, The Center for Informed Choice, The Dartmouth Institute for Health Policy and Clinical Practice

Darleen Mimnaugh, BS, MS - Institute Coordinator
Project Coordinator, The Center for Informed Choice, The Dartmouth Institute for Health Policy and Clinical Practice

Molly Castaldo, JD, MPH – Patient Coordinator
The Center for Informed Choice, The Dartmouth Institute for Health Policy and Clinical Practice

Susan Berg, MS, CGC
Program Coordinator/Interim Director, The Center for Shared Decision Making, Dartmouth-Hitchcock Medical Center

N. Ashley Harris, BS
Project Coordinator, The Center for Informed Choice, The Dartmouth Institute for Health Policy and Clinical Practice

Stephen Kearing, MS
Research Associate, The Center for Informed Choice, The Dartmouth Institute for Health Policy and Clinical Practice

Alyssa Stevens, BS
Administrative Coordinator, The Center for Shared Decision Making, The Dartmouth Institute for Health Policy and Clinical Practice and Dartmouth-Hitchcock Medical Center

Sherry Thornburg, BS, MPH
Project Coordinator, The Center for Informed Choice, The Dartmouth Institute for Health Policy and Clinical Practice

FACULTY ADVISORY COMMITTEE

Kate Clay, MA, BSN, RN
Program Director, Center for Shared Decision Making,
    Dartmouth-Hitchcock Medical Center
Program Director, Office of Professional Education and Outreach,
    The Dartmouth Institute for Health Policy and Clinical Practice
Lebanon, NH, USA

Steve Laitner
General Practitioner & Consultant in Public Health Medicine
Associate Medical Director, NHS East of England; National Clinical Lead for
    Shared Decision Making (Quality and Productivity, Department of Health)
Clinical Advisor to Elective Care and Diagnostics, Department of Health
St. Albans, England, UK

Victor M. Montori, MD, MSc
Professor of Medicine; Consultant, Divisions of Endocrinology and
    Health Care and Policy Research, Departments of Medicine and
    Health Sciences Research
Director, Healthcare Delivery Research Program
Director, Mayo Clinic Share Decision Making National Resource Center,
    Mayo Clinic
Rochester, MN, USA

Lawrence Morrissey, MD
Medical Director of Quality Improvement, Stillwater Medical Group
Stillwater, MN, USA

Richard Thomson, BA, MD, FRCP, FFPHM
Professor of Epidemiology and Public Health, Institute of Health and Society,
    Medical School Newcastle University
Newcastle, England, UK

Craig Westling, MS, MPH
Managing Director, Office of Professional Education and Outreach,
    The Dartmouth Institute for Health Policy and Clinical Practice
Lebanon, NH, USA

Richard Wexler, MD
Director, Patient Support Strategies, The Informed Medical Decisions Foundation
Boston, MA

APPENDIX IV

Summer Institute Participants, 2011

Nerissa Bauer, MD, MPH
Indiana University School of Medicine, Indianapolis IN, USA

Geri Lynn Baumblatt, MA
Emmi Solutions, Chicago IL, USA

Annette Beasley, CNS
Cardiff and Vale University Health Board, Cardiff, Wales

Susan Berg, MS
Dartmouth-Hitchcock Medical Center, Lebanon NH, USA

Bettina Berger, DrPH
University of Witten-Herdecke, Herdecke, Germany

Licia Berry-Berard, MSW
Dartmouth-Hitchcock Medical Center, Lebanon NH, USA

Ruth Z. Bleyler
Patient Advisor, Hanover NH, USA

Laura Boland, MSc, S-LP(C)
Children's Hospital of Eastern Ontario, Ottawa ON, Canada

Charles D. Brackett, MD, MPH
Dartmouth-Hitchcock Medical Center, Lebanon NH, USA

Andrew Carson-Stevens, BSc, MB, BCh, MPhil
Cardiff University, Cardiff, Wales

Molly Ganger Castaldo, JD, MPH
The Dartmouth Institute for Health Policy and Clinical Practice,
    Hanover NH, USA

Kathleen Clark, PhD, JD, MAM
Servant Lawyership, Pleasant Hill CA, USA

Kate Clay, MA, BSN, RN
The Dartmouth Institute for Health Policy and Clinical Practice,
    Lebanon NH, USA

Nicolas Couet, MA, BPhil
Canada Research Chair in Implementation of Shared Decision Making
    in Primary Care, Canada

Cynthia Crutchfield, MBA
The Dartmouth Institute for Health Policy and Clinical Practice,
    Lebanon NH, USA

Dave deBronkart
Society for Participatory Medicine, Nashua NH, USA

Amy Dressler, M.B.A., M.Ed.
Patient Advisor, Hanover NH, USA

Carol F. Dubois
Patient Advisor, Hanover NH, USA

Marie-Anne Durand, BSc, MSc, MPhil, PhD
National Health Service East of England, Cambridge, England

Glyn Elwyn, BA MB BCh MSc FRCGP PhD
Cardiff University, Cardiff, Wales

L. J. (Lyle J) Fagnan, MD
Oregon Rural Practice-Based Research Network, Portland OR, USA

Maureen Fallon
Cardiff and Vale University Health Board, Cardiff, Wales

Audrey Ferron Parayre, MMSc
Laval University Medical Research Center, Quebec City Quebec, Canada

Janice Fischel
Patient Advisor, Hanover NH, USA

Nicole Fowler, PhD
University of Pittsburgh, Pittsburgh PA, USA

Dominick Frosch, PhD
Palo Alto Medical Foundation, San Francisco CA, USA

Melanie Gaiser, MPH, MA
Brandeis University, Waltham MA, USA

Paul E Garfinkel, MHS, CIP
Nemours, Jacksonville FL, USA

Meg Gassert, BA
The Informed Medical Decisions Foundation, Boston MA, USA

Nancy Coffey Heffernan
Patient Advisor, Hanover NH, USA

Jody Hoffer Gittell, PhD, MA
Brandeis University, Waltham MA, USA

Marjorie Godfrey, PhD(c), RN
The Dartmouth Institute for Health Policy and Clinical Practice, Lebanon NH, USA

Ellen Goldstein, MD
Dartmouth-Hitchcock Medical Center, Lebanon NH, USA

Nora-Ashley Harris, BS
The Dartmouth Institute for Health Policy and Clinical Practice, Lebanon NH, USA

Allison Hawke, BA
The Dartmouth Institute for Health Policy and Clinical Practice, Lebanon NH, USA

Susan Hrisos, BSc
Newcastle University, Newcastle Upon Tyne, England

Clarissa Hsu, PhD
Group Health Research Institute, Seattle WA, USA

Natalie Joseph Williams, BSc GDipPsyc
Cardiff University, Cardiff, Wales

Stephen Kearing, MS
The Dartmouth Institute for Health Policy and Clinical Practice, Lebanon NH, USA

Don Kemper, MPH
Founder and CEO Healthwise Inc., Boise ID, USA

Courtney Kozloski
Dartmouth-Hitchcock Medical Center, Lebanon NH, USA

Steven Z Kussin, MD
Shared Decision Center, Utica NY, USA

Steve Laitner, GP
National Health Service East of England, Fulbourn, England

Joanne Lally, MSc, BA
Newcastle upon Tyne NHS Foundation Trust, Newcastle, England

Margaret L Lawson, MD, MHSc
Children's Hospital of Eastern Ontario, Ottawa ON, Canada

Lauren Leavitt, MA
Massachusetts General Hospital, Boston MA, USA

Laurel K Leslie, MD, MPH
Tufts Medical Center, Boston MA, USA

Carmen Lewis, MD, MPH
University of North Carolina, Chapel Hill NC, USA

Ellen Lipstein, MD, MPH
Children's Hospital Medical Center, Cincinnati OH, USA

Sheila Macphail, MBBS, Ph.D.
Newcastle Hospitals NHS Foundation Trust, Newcastle Upon Tyne, England

Irma H Mahone, RN, Ph.D.
University of Virginia School of Nursing, Charlottesville VA, USA

Melanie Mastanduno, BSN, MPH
The Dartmouth Institute for Health Policy and Clinical Practice,
    Lebanon NH, USA

Dan D. Matlock, MD, MPH
University of Colorado School of Medicine, Aurora CO, USA

Suepattra May, MPH, PhD
Palo Alto Medical Foundation Research Institute, Palo Alto CA, USA

Helen McGarrigle
Cardiff and Vale Breast Unit, Cardiff, Wales

Douglas C. McKell, MS
Connecticut Surgical Group, Hartford CT, USA

Katherine Milligan, MBA, PhD
Tuck School of Business, Hanover NH, USA

David Milne
Patient Advisor, Hanover NH, USA

Darleen Mimnaugh, BA, MS
The Dartmouth Institute for Health Policy and Clinical Practice, Lebanon NH, USA

Lisa A Mistler, MD, MS
University of Massachusetts Medical School, Worcester MA, USA

Larry Morrissey, MD
Stillwater Medical Group, Stillwater MN, USA

Albert G. Mulley, Jr, MD, MPP
The Dartmouth Center for Health Care Delivery Science, Hanover NH, USA

Mary Ann Murray, PhD, RN
The Ottawa Hospital, Ottawa ON, Canada

Catherine Nadeau, MSc (c)
Laval University Medical Research Center, Ste-Foy QC, Canada

Claire Neely, MD
Institution For Clinical System Improvements, Bloomington MN, USA

Edmund Alwin Martin Neugebauer, PhD
University of Witten Herdecke, Cologne, Germany

Alan David Nye, MBBCh
National Health Service - NHS Direct, England

Elissa Ozanne, PhD, MS
University of California San Francisco, San Francisco CA, USA

Lia Palileo, MD, MSc
University of the Philippines, Quezon City, Philippines

Antony Porcino, PhD(c)
CAMEO-Research Program, Vancouver BC, Canada

Roy Proujansky, MD, MBA
Nemours du Pont Hospital for Children, Wilmington DE, USA

Eva M. Rzucidlo, MD, FACS
Dartmouth-Hitchcock Medical Center, Lebanon NH, USA

Mary E. Ropka, PhD, RN
University of Virginia School of Medicine, Charlottesville VA, USA

Janet Schuerman, MBA
Institution For Clinical System Improvements, Bloomington MN, USA

Karen Sepucha, PhD
Massachusetts General Hospital, Boston MA, USA

Corey A. Siegel, MD
Dartmouth-Hitchcock Medical Center, Lebanon NH, USA

Leigh Simmons, MD
Massachusetts General Hospital, Boston MA, USA

Rupert Sobotta, MA, MD
Gemeinschaftskrankenhaus Herdecke, Herdecke, Germany

Mike Spencer
Cardiff and Vale University Health Board, Cardiff, Wales

Jodi Sperber, MPH, MSW
Brandeis University, Waltham MA, USA

Simone Steinhausen, MSc
University Witten Herdecke, Koeln, Germany

Alyssa Stevens, BS
Dartmouth-Hitchcock Medical Center, Lebanon NH, USA

Elizabeth Teisberg, PhD
University of Virginia, Charlottesville VA, USA

NC Tenenbaum, BA
MedicalFrontiers.com, Mount Kisco NY, USA

Richard Thomson, MD, FRCP, FFPHM
Institute of Health & Society - Newcastle University,
    Newcastle Upon Tyne, England

Sherry Thornburg, BS, MPH
The Dartmouth Institute for Health Policy and Clinical Practice,
    Lebanon NH, USA

Caroline Tietbohl, BA
Palo Alto Medical Foundation, Palo Alto CA, USA

Dale Collins Vidal, MD, MS
Dartmouth-Hitchcock Medical Center, Lebanon NH, USA

Shelley Volz, MA
University of California at San Francisco, San Francisco CA, USA

Amy Tu Wang, MD
Mayo Clinic, Rochester MN, USA

James Weinstein, DO, MS
Dartmouth-Hitchcock Medical Center, Lebanon NH, USA

Richard Wexler, MD
The Informed Medical Decisions Foundation, Boston MA, USA

Renda Soylemez Wiener, MD, MPH
Boston University School of Medicine, Boston MA, USA

Linda Wilkinson, MBA
Dartmouth-Hitchcock Medical Center, Lebanon NH, USA

Tim Wysocki, PhD
Nemours Children's Clinic, Jacksonville FL, USA

Jingyu Zhang, PhD
Philips Research North America, Briarcliff Manor NY, USA